GREEK SCIENCE IN ANTIQUITY

Greek Science in Antiquity

MARSHALL CLAGETT

COLLIER BOOKS
A Division of Macmillan Publishing Co., Inc.
NEW YORK
COLLIER MACMILLAN PUBLISHERS
London

© 1955 by Marshall Clagett
All rights reserved. No part of this book may be reproduced or transmitted in any form or by any means, electronic or mechanical, including photocopying, recording or by any information storage and retrieval system, without permission in writing from the Publisher.
First Collier Books Edition 1963
Sixth Printing 1979
This edition, revised for Collier Books, is published by arrangement with Abelard-Schuman Limited
Macmillan Publishing Co., Inc.
866 Third Avenue, New York, N.Y. 10022
Collier Macmillan Canada, Ltd.

PRINTED IN THE UNITED STATES OF AMERICA

To

LYNN THORNDIKE

Preface

I HAVE WRITTEN this volume for the adult or student reader without specialized scientific background, but with a knowledge of secondary-school mathematics. The reader I have visualized, then, is the nonspecialist who is beginning the study of history of science, although I hope the work will not be without interest for the specialist as well.

In this volume I have attempted to give especial and marked attention to the fate of Greek science in late antiquity. Elementary texts in the past have long ignored this aspect of Greek science. The importance of the course of Greek science in late antiquity is evident, for it was during this period that much of the Greek scientific corpus was put into the form in which it passed to the medieval Latin West. We are justified, then, in considering this volume as an introduction to medieval and early modern science—that science being considered as a transformation of Greek science. In fact, I am preparing a volume on medieval science that will take up the development of science where this volume leaves off.

One of my colleagues has raised a question as to whether the title *Greek Science in Antiquity* is a tautology. My answer is no for two important reasons. In the first place there is other than Greek Science in antiquity (e.g., Babylonian Science or Indian Science). Secondly and more important, I have stressed throughout the volume that the Greek scientific corpus has had a long life in antiquity, in the Middle Ages, and in early modern times. This volume covers only the first part of that long life.

I should like to give special thanks to Professor Thomas Smith, of the California Institute of Technology, who for several years assisted me in the adventure of trying to teach Greek science to nonspecialized students, and who read with care the first draft of this book. Thanks are due also to Professor I. E.

Drabkin, of the City College of New York, who read the manuscript in typescript and offered a number of valuable suggestions for its improvement, and to my colleague Professor R. C. Stauffer of the History of Science Department, who read proof and saved this volume from a number of errors.

M. C.

University of Wisconsin

Preface to the Second Edition

I HAVE CORRECTED what seemed to me as glaring errors in the first edition. One is always tempted to rewrite a volume after so many years. I have, however, resisted that temptation. Instead of adding a section on Stoic physics or substantially rewriting the section on the mathematical side of Greek astronomy, I have given references to the excellent treatments of Sambursky (for the Stoics) and Neugebauer (for astronomy) where the reader will find lucid explications which have not passed through my debilitating sieve. I particularly want to thank Professor O. Neugebauer of Brown University for suggesting some crucial emendations to my brief treatment of Greek astronomy. I also want to thank Professor William Stahl of Brooklyn College for his suggestions on late Latin science in a review of the first edition. Finally, I must thank my colleague Professor William Stahlman for his helpful comments in connection with Chapter 7.

Additions have been made to the section of further readings, although no attempt has been made to cull all of the literature that has appeared since the first edition. This is particularly true of the second section on late antique science and philosophy.

M. C.

University of Wisconsin, 1962

Contents

Part One / The Origins of Science in Antiquity

1	Science in Egypt and Mesopotamia	13
2	Greek Science: Origins and Methods	33
3	Science and Early Natural Philosophy	49
4	Greek Medicine and Biology	54
5	Greek Mathematics	71
6	Greek Physics	84
7	Greek Astronomy	106
8	Roman Science	127

Part Two / Science in Late Antiquity

9	Science and Spiritual Forces in Late Antiquity	143
10	Science and Patristic Literature	161
11	Latin Science in Late Antiquity	180
12	Latin Science in the Early Middle Ages	193
13	Greek Science in the Age of Justinian	206
14	In the End the Beginning	224

Appendices

1	Archimedes and the Quadrature of the Circle	227
2	Archimedes and the Application of Mechanics to Geometry	229
3	Apollonius of Perga and the Development of Conic Sections	231
4	Archimedes' Proof of the Law of the Lever	236
5	Ptolemy's Table of Chord Lengths	238

Further Readings 245

Index 252

PART ONE

The Origins of Science in Antiquity

Chapter 1

Science in Egypt and Mesopotamia

I

TODAY MOST of us are keenly aware of the close ties that exist between science and the character of our civilization. The historian of science hopes to throw light on this relationship by investigating how and why science grew out of the more general province of learning to occupy its dominant position. In this investigation he must study the unfolding life of science. He must, in addition, evaluate the details uncovered and judge their significance for modern scientific attitudes, principles, and methods.

As in the history of any other aspect of man's activity—politics, economics, religion, etc.—so in the history of his science, a study of the early stages of growth is of great importance for an understanding of the modern institution. This volume attempts to treat the phases of science that lie in antiquity—from the very early empirical efforts of prehistoric man through the relatively mature scientific activities in Greek antiquity. Our volume will, then, cover an enormous period of time, of which the beginning is unrecorded and the terminal date is about A.D. 600.

It is generally agreed that the learning of antiquity was digested in the Middle Ages and Renaissance to form the chief nourishment for growth of early modern thought. Thus early modern science grew out of Greek science and philosophy as modified by the natural philosophers of Islam and the Latin West. The acceptance of this essential continuity in the development of Western thought does not obviate the novelty of the scientific activity of the seventeenth century, the century of Galileo, Boyle, Hooke, Leibniz, and Newton. It serves rather to clarify that novelty, to show how it arose in great part from the interplay, modification, and rearrangement of older stock ideas as they were fashioned into an essentially new system.

In the course of our inquiry into the foundations of science we must direct our attention toward many aspects and features of intellectual activity so that we may achieve a relatively complete picture. Special care must be exercised in the exposition and evaluation of the more important scientific principles, laws, and theories that matured, suffered modification, or originated in antiquity. But the mere cataloguing of the substantial scientific knowledge of antiquity would ill satisfy our objective. We must also turn our attention to the methods and procedures pursuant to that science. We must show the relative importance of revelation, authority, reason, experience, and experimentation, and how the emphasis on one or another of these factors changed. And we must examine the more general philosophical ideas that influenced both the method and the substance of science in antiquity, particularly during the beginnings of Greek science.

In setting out to characterize this young, active, and exciting period in the life of science, we should, no doubt, state our definition of science, although we are tempted, like one recent historian of science, to beg the question until the history has been written and the reader can judge for himself what kind of enterprise has been described. The difficulty is that there is little agreement on the question of definition among the professional scientists or among historians and philosophers of science. And yet most of us have a rather intuitive understanding of the province of science and to some extent of its methods. Schoolboys speak easily of laboratories with betatrons, of fission, of penicillin, of frequency modulation, of the expanding universe, and of a myriad of other scientific topics. It will nevertheless be useful to go beyond our ill-defined intuition of science to accept some definition, however arbitrary, as a point of departure for this study.

For the sake of breadth we would be inclined to accept the definition of one of the pioneer historians of science, George Sarton, who defines science as systematized positive knowledge, were it not for the fact that the term "positive" is ambiguous and demands commentary. Also, this definition would seem to deny or at least to exclude important nonpositivistic elements that were once closely united with science—e.g., magic. We can, perhaps, avoid these difficulties and still retain sufficient breadth by assuming that science comprises, first, the orderly and systematic comprehension, description, and/or explanation of natural phenomena and, secondly, the tools necessary for that undertaking. The first part of the definition permits us to inquire into man's scientific activity even in times of which we have no written records. The latter part of the definition allows us to treat the growth of mathematics and logic as a part of the history of science.

II

Science as an orderly and rational structure scarcely goes beyond the Greeks into the early history of man. Yet we should be aware of the countless centuries of man's activity

FIG. 1. The Celtic horse and the reindeer, from engraved outlines at the Grotte de la Mairie, in Dordogne, France.

before the inception of written records, as well as of the two thousand years or so of "civilized" activity before the emergence of Greek culture.

In prehistory—before the invention of writing—Stone Age man left us many evidences of his understanding of nature. Who cannot feel the eye for naturalistic detail revealed in the Stone Age wall paintings and reliefs in the caves of southern France and northern Spain (see Fig. 1)? Man was a naturalist long before he was a farmer, an architect, or a politician. At the same time, there can be little doubt that in the manufacture of his tools early man made a permanent record of his experience with and understanding of nature. This idea has been developed in an interesting, if somewhat extravagant, way by the British archeologist V. Gordon Childe in various of his publications. Childe sees the tool as "an embodiment of science . . . a practical application of remembered, compared, and collected experience of the same kind as are summarized in scientific formulae, descriptions, and prescriptions." Although this statement glosses too readily over the generalization and abstraction involved in scientific formulae and description, there is a kernel of truth in its representation of the tool as a product of man's understanding of nature.

In studying the activity of early man, Childe and other historical materialists have done important service by emphasizing the influence of man's technological activities upon both his general cultural development and his particular scientific advances. The control and use of fire was one of his earliest cultural discoveries and certainly one of the most significant. The chemical and metallurgical arts stem from that fundamental discovery and the complementary discovery that charred wood ignites more easily than uncharred. Thus early man had the chief instrument for learning the protochemistry involved in the firing and burnishing of pots and in the working of metals.

Man emerged into civilization in the fourth millennium B.C. in at least three important areas: the Nile Valley, Mesopotamia, and the Indus Valley. This emergence was due in part to his discovery of the use and working of metals. With the coming of the metal ages his practical knowledge of natural phenomena multiplied. He learned to reduce and work copper, to make use of its ready malleability and fusibility, to alloy it with tin to produce bronze, to make castings by delicate and sure processes. He also made brick arches, caught the wind in sail, and applied the wheel to pot and vehicle. He learned to do all this, no doubt, by a long and arduous trial-and-error method—or, we may say, by experimentation, so long as we are careful to distinguish the experimentation involved in the discovery and improvement of the technological arts from the planned experimentation used for the confirmation of scientific theory. Finally, with the invention of writing, possibly first in the area of Mesopotamia about the middle of the fourth millennium B.C., man stepped over the threshold of civilization. This invention has been described by some authors as a social accident tied in with the development of ownership signs and property lists. It has, of course, been the chief instrument in the perfection of man's cultural and thus of his scientific achievement.

III

The study of rationally organized scientific activity, as distinguished from handicraft activity, in two of the first areas of civilization, Egypt and Mesopotamia, reveals advances primarily in mathematics and astronomy, and in some aspects of medical practice. The rest of man's understanding and description of nature remained either entirely in the state of technology, as in the case of his knowledge of chemical and mechanical principles, or in that of mythology and magic, as in the case of his cosmology and much of his medicine.

Just as the approach of the early inhabitants of Egypt and Mesopotamia toward the origin of the world and man was almost entirely mythological and religious, and just as their reliance for the explanation of natural phenomena was upon supernatural causation, so much of the medical practice in Egypt and Babylonia was based upon belief in other than natural causes. There was considerable dependence upon incantations and other magical cures. For example, one of the medical papyri from Egypt, the so-called Ebers Papyrus, compiled in the eighteenth dynasty (about 1500 B.C.), presents considerable evidence that the sorcerer worked side by side with the ordinary practitioner (and, no doubt, was often the same person), utilizing spells and incantations for desired therapeutic ends. Thus, in applying malachite for a cataract of the eye, we are told to incant the following, after having pounded malachite with honey: "Come Malachite. Come thou green one; Come discharge from Horus' eye. Come secretion from Atum's eye. Come fluid that has come out of Osiris. . . ." G. Conteneau, the most able historian of medicine in the Mesopotamian area, shows how thoroughly bound to the whim of gods was the Assyrian-Babylonian pathology and therapeutics. Medical treatment in numerous cases took the form of mollifying the gods by prayer and sacrifice and expelling demons by incantations and medicines.

But there is one medical treatise from early antiquity that makes us hesitate in dismissing Egyptian medicine, at least, as completely magic-perverted. This is the renowned Edwin Smith Surgical Papyrus. Named for its modern purchaser, it was copied in the seventeenth century B.C. from a work of the third millennium B.C. Free of the magical features of other Egyptian medical works, it is organized in a systematic manner and approaches a scientific presentation of empirical data. Although incomplete, it contains forty-eight cases involving injuries and wounds to various parts of the body. It starts with cases occurring to the head—to the nose, face, and ears—and then proceeds to the collarbone, humerus, thorax, shoulder, and spinal column. The presentation of each case is divided into a title, an examination, a diagnosis and opinion, a treatment, and glosses to explain already antique and specialized terminology.

Almost as remarkable as the quasi-scientific organization of the treatise is its clear indication of the straightforward simplicity and restraint practiced in the cures. Particularly noteworthy is the admission in some thirteen cases that the wound is fatal—i.e., it is an "ailment not to be treated." J. H. Breasted, the modern editor and translator of the papyrus, in his edition (which will remain one of the great monuments of careful and complete scholarship in the fields of Egyptology and the history of medicine) justly remarks on the significance of describing cases not treatable.

> The inclusion of these thirteen hopeless cases (about one fourth of the material in the treatise as preserved to us) is a remarkable evidence of the surgeon's scientific interest in recording and discussing the observable facts in a group of cases for which he could suggest no treatment. (J. H.

Breasted, *The Edwin Smith Surgical Papyrus,* Vol. 1, Chicago, 1930, p. 47.[1])

It has been suggested that the inclusion of hopeless cases demonstrates less scientific prudence than a reluctance to face the social, or perhaps even legal, reprisals against unsuccessful diagnosis and treatment.

The factual anatomical knowledge exhibited by Egyptian medical writings is not extensive. But the Edwin Smith papyrus does give us the first, although incomplete, description of the brain, and it says something of the brain as a control center for the nervous system. It also takes note of the use of the pulse "to know the action of the heart," and it reveals a rudimentary knowledge of the blood system, although there is no evidence of an understanding of the circulation of the blood. Citation here of Case 6 will give the reader a vivid picture of both the organization of and the state of the substantial surgical knowledge in the treatise.

Case 6

(Title) Instructions concerning a gaping wound in his head, penetrating to the bone, smashing his skull, (and) rending open the brain of his skull.

(Examination) If thou examinest a man having a gaping wound in his head, penetrating to the bone, smashing his skull, and rending open the brain of his skull, thou shouldst palpate [i.e., feel with your fingers] his wound. Shouldst thou find that smash which is in his skull [like] those corrugations which form in molten copper, (and) something therein throbbing (and) fluttering under thy fingers, like the weak place of an infant's crown before it becomes whole . . . (and) he [the patient] discharges blood from both his nostrils, (and) he suffers with stiffness in his neck,

(Diagnosis) [you say] "an ailment not to be treated."

(Treatment) [But] thou shouldst anoint that wound with grease. Thou shalt not bind it; thou shalt not apply two

[1] This and the succeeding quotation are quoted by permission of the publisher, the University of Chicago Press; copyright, 1930, by the University of Chicago.

strips upon it until thou knowest he has reached a decisive point.

(Gloss A, Explaining:) "Smashing his skull and rending open the brain of his skull," (it means) the smash is large, opening to the interior of his skull, (to) the membrane enveloping his brain, so that it breaks open his fluid in the interior of his head.

(Gloss B, Explaining:) "Those corrugations which form in molten copper:" it means copper which the coppersmith pours off (rejects) before it is [forced] into the [mold], because of something foreign upon it like [wrinkles]. It is said: "It is like ripples of pus." (*Ibid.*, pp. 164-173.)

IV

These rudimentary efforts to organize empirical data in a scientific manner in surgery were perhaps less influential in the further development of science than the prescientific efforts exerted in astronomy and mathematics by the peoples of Mesopotamia and Egypt. With certain exceptions in the mathematics and astronomy of Mesopotamia to which we shall briefly allude in a moment, these preliminary steps toward science taken in pre-Greek antiquity have left imprints of a protoscience largely empirical in character—i.e., largely observational, with little regard for theory.

One of the principal products of the empirical procedure in astronomy in both Egypt and Mesopotamia was the development of more accurate calendars. Motivating early calendaric development was a variety of social needs: the demands of political administration, such as the need for the establishment of regular tax-collecting days; the pressing requirements of agriculture, including the need for advance determination of flood seasons and planting times; the fixing of regularly occurring religious feast days; and so on. The rise and concurrent use of a number of different calendars in both Egypt and Mesopotamia has left our modern study of them in considerable confusion. But it is known that in Egypt an official calendar of 365 days (twelve months of thirty days each plus five additional days) was utilized during much of the historical period. This has been called the "sliding" or "wandering" year; since it was almost one fourth of a day shorter than the tropical year of the seasons, its New Year's Day came about one fourth of a day earlier in the tropical calendar each year. Thus in a period close to 1460 tropical years the New Year's Day of the sliding year slid through every day of the tropical year.

Existing side by side with the official calendar in Egypt was a lunar calendar of 354 days and twelve months which, recent research suggests, extended from the first day after the old crescent to the first day after the next old crescent, in contradistinction to the months of most lunar calendars, which extended from one new moon to the next. Apparently a simple empirical rule was used to add on occasion a thirteenth month to bring the year in line with the official year, and it was perhaps not until the fourth century B.C. that a schematic lunar calendar based on a predicted cycle was introduced.

It has long been supposed that the fact that the official calendar, with its "wandering" year, would in time fail to correspond with the seasons led the Egyptians to invent a fixed sidereal year quite close to the tropical year. It was believed by scholars that the New Year's Day of this sidereal year of approximately 365¼ days was determined by observing when the bright star Sirius (Sothis) first rose just before dawn after its long period of invisibility. But the scanty evidence for such a Sothic fixed year is most ambiguous, and at least one recent investigator asserts rather that the rising of Sirius was used to bring a lunar calendar into some agreement with the seasons. Thus, according to this view, it was the lunar calendar that served as an agricultural or seasonal calendar.

In the Mesopotamian area it was the lunar calendar that received official support throughout the historical period. Since a calendar of 354 days was about 11¼ days short of the tropical year of the seasons and so would soon be out of adjustment with the seasons and of little use for agricultural activities, the practice arose, as in Egypt, of adding an extra month in certain years. The technical term used for this practice is "intercalation." This intercalation of an extra month was at first an empirical rather than a systematic act, but in the late historical period, possibly by 480 B.C. and probably by the fourth century B.C., a regular system of intercalation, known among the Greeks as the Metonic cycle, was in effect. Each nineteen-year cycle contained seven leap years of thirteen months, with one particular month being repeated in six of the years and another month in the seventh.

The constant effort to coordinate the lunar and the tropical years seems to have been one of the important stimuli in this area to careful and continual observation of the apparent movements of the moon and the sun. This activity became "scientifically significant" at least as early as 747 B.C., the date of the first regularly repeated observations under the Assyrians. These observations were recorded in the form of tables which were apparently used by the later Greek astronomers and which were a great tribute to the patience and skill of Babylonian astronomers of Mesopotamia, although the length and completeness of these observations appear to have been greatly exaggerated by modern authors. Many pertinent scientific data were organized in these tables. It is immaterial in the progress of science whether the motivation to this observation was solely calendar-fixing or was in part astrological (as it was if we assay correctly the significance of a number of astrological tablets that date all the way back to the third millennium B.C., although here again the significance of the astrological material for the development of mathematical astronomy seems to have been unduly stressed by modern investigators). In either case the result was to stimulate the application of mathematics to the observational data for the purpose of predicting the future positions of the sun, the moon, and, on occasion, the planets. This was an extraordinary step, for it is the application of mathematics to observational data which lies at the heart of the method of modern science.

One of the foremost students of ancient astronomy, O. Neugebauer, has brilliantly described this application of mathematics to astronomy, showing how it led to the development of a very important technique in astronomy. He points out how there developed from about 700 B.C. preliminary methods of prediction based on long-range observations. By these methods over-all averages for the main periods of astronomical phenomena could be obtained. These averages then might be improved by occasional individual observations. At the same time, short-range predictions of phenomena could be made on the basis of a series of observations immediately preceding the event.

> After such methods had been developed to a certain height, apparently one ingenious man conceived a new idea which rapidly led to a systematic method of long-range prediction. This idea is familiar to every modern scientist; it consists in considering a complicated periodic phenomenon as the result of a number of periodic effects, each of a character which is simpler than the actual phenomenon. The whole method probably originated in the theory of the moon, which we find at its highest perfection. (O. Neugebauer, "The History of Ancient Astronomy: Problems and Methods," *Journal of Near Eastern Studies*, Vol. 4 [1945], p. 9.)

The basic problem was to predict the first visibility of the crescent each month, for the beginning of the month was established on the basis of this phenomenon. The time between the successive first visibilities is never more than thirty days and never less than twenty-nine days, but the fundamental question is: which of these alternatives is it for each month? Now this complicated periodic phenomenon—namely, the first visibility of the crescent—is influenced by a number of factors that have independent variations of their own: the moon's velocity, the sun's "velocity," the length of daylight for the point of observation, the moon's movement in latitude, etc. It was one of the great glories of Babylonian astronomy to recognize that this complicated periodic phenomenon was the result of a series of independent periodic variations. When we examine the lunar ephemeris in detail—column by column—we realize that each of the variables is being considered and its contribution to the general phenomenon noted. And, finally, one of the most important features of the ephemerides, which appeals to the modern reader with some astronomical and mathematical training, is the way in which linear methods were used to approximate complex functions. The Babylonian astronomers used arithmetical series in which the terms uniformly increased to a maximum and then uniformly decreased to a minimum; and we can find by extrapolation the crucial maximal and minimal points they used. It is worth noting that the Babylonian procedures seem to have been entirely arithmetical. Unlike the Greeks, the Babylonians did not assume geometrical models.

Before concluding our survey of the initial steps taken in scientific astronomy among the peoples of Egypt and Mesopotamia, we must note in passing that this pre-Greek period gave us not only the long-range time system of the calendars but also the first short-term timepieces. A variety of sundials and water clocks were devised in pre-Greek times which contained the essential features of timepieces used until the rise of mechanical clocks in the high and late Middle Ages. It is worth remembering that for accurate time measurement in a scientific experiment Galileo, in the seventeenth century, resorted to a water clock, a product of the first civilizations, rather than to the then inefficient mechanical clock, which was in the next two centuries to become the most essential of scientific instruments. It is also worth noting that the measurement of time was of scientific moment in antiquity and the Middle Ages only in astronomy. In physics it had to wait on the experimental investigation of mechanics.

V

It would seem from our brief account that Babylonian astronomy developed further than did that of Egypt. Why was this so? The instruments of the Babylonian astronomer were probably not superior to those of the Egyptian astronomer, for both possessed only the simplest sighting instruments. Nor was the system of constellations or star arrangements devised by the Babylonian astronomer to act as a reference frame on which he might plot his data markedly superior to that of the Egyptian astronomer, at least in the earlier period. Part of the answer to this question lies in a fact that has already been suggested, that in Babylonia the lunar calendar was adopted and continually had to be adjusted to the seasons, thus demanding detailed calculations of the moon's movements. But it is likely that an equally persuasive reason for the superiority of Babylonian astronomy lies in the development of a superior system of numerical calculations. Babylonian mathematics seems to have advanced considerably beyond that of Egypt —except perhaps in practical geometry.

A glance at the mathematical papyri of Egypt will readily convince us of the grossly empirical nature of Egyptian mathematics and will reveal the awkward and cumbersome nature of the system of numeration. The system was both "decimal" and "additive." It was decimal in that a new symbol was employed for each power of ten. Thus there were distinct symbols for 1, 10, 100, 1000, . . . 1,000,000. The system was without any concept of "place-value," which, we shall see shortly, was developed in Mesopotamia.

The additive nature of Egyptian numeration is immediately evident in their methods of calculation. Thus in the multiplication, shown below, of 5 × 7 this arrangement would be employed.

√1	√7
2	14
√4	√28
5	35

FIG. 2.

The first step is to take 7 once, then to double it, and then to double the duplication. After successive duplications we add from the column of multipliers the numbers 1 and 4 to make the desired multiplier 5. Thereupon we add up the numbers opposite the multipliers 1 and 4—namely, 7 and 28. This gives us for the multiplier 5 the final product 35. Hence by combining addition with the simplest possible multiplication—i.e., duplication, itself akin to addition—a more extended multiplication was accomplished. With larger numbers multiplication by ten was used as well as duplication. A similar procedure was used in division.

The cumbersome and empirical nature of Egyptian mathematics is further illustrated by the concept and treatment of fractions found in the celebrated Rhind Mathematical Papyrus, copied about 1660 B.C. from an original of about 1849-1801 B.C. Fractions for the most part were reduced to fractions with the numerator of one. Utilized for this reduction was the so-called "table of two" found in the papyrus. This table gives us the reduction of fractions with numerator 2 and odd-numbered denominators up to 101 to fractions with numerator 1. Thus $2/7 = 1/4 + 1/28$. The actual manipulation of fractions was quite complicated, for it seems to have involved a painful trial-and-error procedure. It is just to say that the Egyptian was held back in the development of arithmetic and algebra by his elaborate procedures in calculation.

It is little wonder, therefore, that, unlike the Babylonians, who carried algebra to a remarkable degree of development, the Egyptians were able to solve only simple linear equations and a few quadratic equations of the type $ax^2 = b$. The foregoing symbolization is used for the sake of clarity; if there was symbolization of a sort in early mathematics, it was not of this kind.

The most advanced of the Egyptian mathematical achievements were in geometry. In dealing with the areas of circles the Egyptians took the square of side 8 as equivalent to the circle of diameter 9. This is the same as using a value of pi equal to 3.16. This may be compared with the Babylonian records which show evidence implying at least two different values for π, namely 3 and $3\frac{1}{8}$.

In addition, the Egyptians knew how to determine the areas and volumes of a number of figures; they could find the area of a triangle and a trapezium, the volume of a cylindrical granary and of a frustum of a square pyramid, and perhaps even the area of the surface of a hemisphere (although the last is doubtful). Their proficiency in geometry was certainly fostered by their high development in architectural engineering and surveying, as Herodotus and other Greek authors have long since reminded us (see Figs. 3 and 4). The solutions of the geometrical problems were never given in a completely general form; i.e., reference was not made to *any* triangle having an area equal to one half its altitude times its base. But specific as the problems and their solutions were, a certain generality obviously was implied. The specific problem given was quite clearly meant to be an *example* problem.

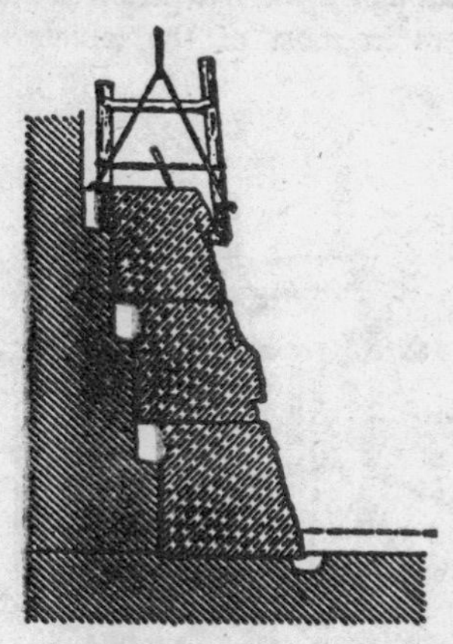

FIG. 3. Method of setting stones in place in pyramid building.

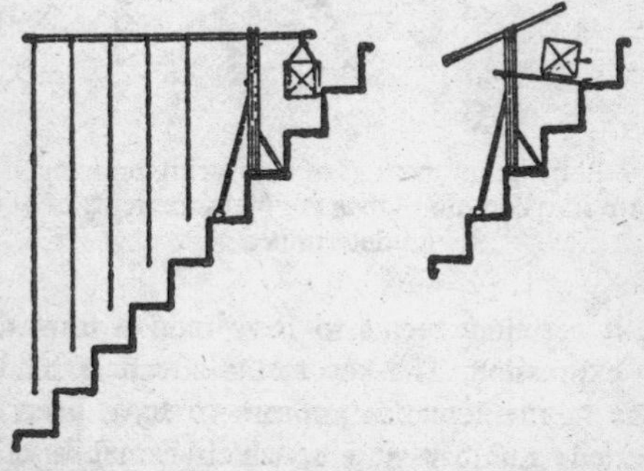

FIG. 4. Diagram showing use of the lever principle to set stones in place.

Presumably the surveyor who had to find the area of a triangular plot of land would use the same procedure that was used in the specific example, changing the numbers to suit his observed values. But we must await the formation of Greek geometry to find explicitly expressed general formulae based on wholly general and abstract concepts of lines, surfaces, and volumes.

When we turn to Mesopotamia, we find from at least 1800 B.C. a Babylonian mathematics more highly developed than the Egyptian. Although it too had strong empirical roots that are clearly present in most of the tablets that have been

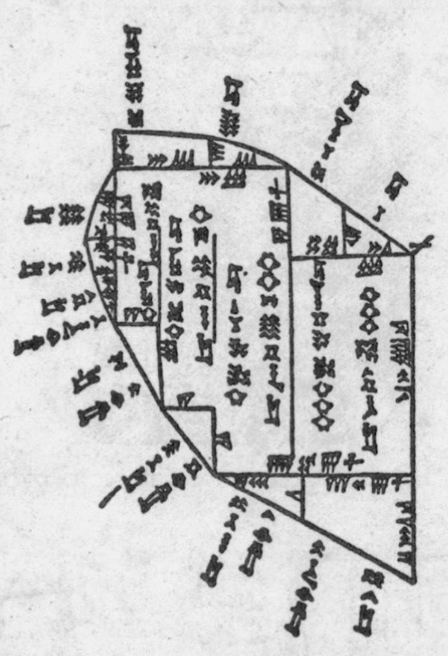

FIG. 5. The Babylonian method of measuring an irregular field by breaking it up into common areas the measurements of which could be determined.

published, it certainly seems to have tended toward a more theoretical expression. The key to the advances made by the Babylonians in mathematics appears to have been their remarkably facile number system, which demands brief characterization.

(1) Although it had certain features of both decimal and sexagesimal systems, it was primarily a sexagesimal system. That is to say, it was based on sixty and powers of sixty.

(2) It was a system highly general and abbreviatory in character. All numbers could be made with only two symbols. 𒁹 = 1 and 𒌋 = 10. Using these symbols the numbers from

1 to 59 can be represented thus: 𒐖 = 2; 𒌋𒌋𒁹 = 21; etc. Numerous tricks were used to save writing all the symbols out in a string. Not only could these symbols be used to represent numbers from 1 to 59 but they could also be used to write the numbers 1 to 59 times any power of 60. Thus unless one knew what order of magnitude was being considered from

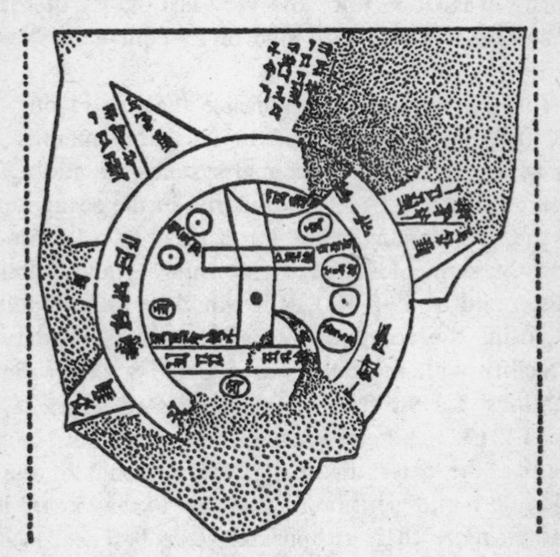

FIG. 6. Babylonian world map.

the details of the problem being worked, he could not know whether the two symbols 𒌋𒌋 by themselves on a tablet without supporting text equaled 20, or 20×60, or 20×60^2, or 20×60^{-1}, etc.

(3) What is more, these same symbols changed their value as their position changed; that is to say, this system was a place-value system, as is our own decimal system. In our system as the symbol 1 changes position in the following numbers it changes its value in that it stands for a higher power of ten: 00.1, 1, 10, 100, etc. So in the number 𒁹𒌋𒁹 the symbol 𒁹 has the value of 1 in the last position and 60 in the first position, the whole number being 71 (if the 𒁹 in the last position represents 60, then the 𒁹 in the first position represents 60^2 and the whole number is 4260: 60^2 plus 11×60).

(4) It was not until very late that the Babylonian system developed what in the decimal system is called zero—i.e., a sign for the absence of any units of a given power of ten indicated by position; thus in our system 101 means, of course, one hundred, no tens, and one unit. Instead of using that sign we could agree that we would simply leave a space, writing 1 1. This was done until the very last stages of the Babylonian system, when a zero sign or its equivalent was developed.

(5) Finally, we should emphasize that, as in our decimal system, fractions were treated in the same manner and as readily as integers. In the decimal systems we add 1.1 to 1.1 in the same way as 11 to 11. Similarly, in the sexagesimal system of the Babylonians it was just as easy to calculate with 1; 1 (using the semicolon to separate integers and fractions, so that this would be $1 + \frac{1}{60}$) as it would be to manipulate 1, 1 (or 61, using the comma to separate powers of sixty). This ready facility with fractions was coupled with fractional approximations, for we find $\sqrt{2}$ approximated as 1; 24, 51, 10 (i.e., 1.414213. . .) in one tablet discovered. As Neugebauer points out, "the determination of the diagonal of the square from its side is sufficient proof that the 'Pythagorean' theorem was known more than a thousand years before Pythagoras."

Freed from the drudgery of calculation by this really remarkable system of calculation, and by the preparation of many kinds of tables (tables of squares, tables of square roots, multiplication tables, reciprocal tables, and others), the Babylonians made extraordinary advances in algebra, in the numerical or algebraic solution of quadratic equations (i.e., equations in which the unknown is squared, such as $x^2 + bx = c$). Apparently, by some procedure similar to making the left side of the equation a perfect square (i.e., by completing the square) and adding the same amount to the right side of the equation, they reached a formula for the solution of such an equation:

$$x = \sqrt{c + \frac{b^2}{4}} - \frac{b}{2}$$

Or perhaps it would be better to say that the Babylonians

solved specific problems involving the type of equation $x^2 + bx = c$ by a series of numerical steps corresponding to the solution we have given in modern symbols above. They solved numerous other problems involving quadratic equations of somewhat different forms, modifying their solutions accordingly. Now it must not be thought that the Babylonians had a general solution of the quadratic equation, couched in general terms. As in the case of Egyptian geometry, specific problems were given to illustrate the main types of numerical procedure in the solution of equations of somewhat different forms. But even without the general solution, the Babylonians laid an extraordinary base for the subsequent development of algebra, and it is difficult—indeed, almost impossible—to believe that the Greek algebraist Diophantus and the Indian and Islamic algebraists do not owe their fundamental methods to these people of Mesopotamia, who worked out the basic solutions of elementary algebra in the second millennium before Christ.

VI

We can now take stock of and assay the essential contributions and features of the science of the pre-Greek period.

(1) This science was above all empirical in nature, without attention to theory, the chief exception being in very late developments in Mesopotamian astronomy and some developments fairly early in Babylonian mathematics. The empirical nature of much of Egyptian mathematics and medicine is evident. A history of the science of this period clearly confirms the empirical origin of science.

(2) The objectives of this early science were largely social. Pure science was almost indistinguishable from applied science. We have noted the important stimulus of the calendaric development to astronomy and of engineering to geometry.

(3) Preliminary steps were taken to organize empirical data in a scientific manner, as reference to the Edwin Smith Surgical Papyrus and to the astronomical tables of Babylon readily confirms.

(4) There were occasional examples in this pre-Greek period of attempts to predict the course of phenomena on the

basis of scientifically organized data. The prediction of the fatal outcome of certain wounds by the author of the surgical papyrus is a rather obvious example; the attempts to predict the positions of the heavenly bodies on the basis of observations constitute an even better example.

(5) And coupled with the attempts at prediction there was strikingly evident in at least one case, Babylonian astronomy, an attempt to treat the description of natural phenomena mathematically. It was also in the field of astronomy that the Greeks were to use mathematics most brilliantly.

(6) It was among the Babylonians that the only theoretical efforts in mathematics developed. However firmly the early mathematics might have been rooted in the practice of the engineer and the administrator, the scribe who compiled the problem texts with algebraic problems and their solutions was already shaking off the dust of practice.

(7) But the sciences of astronomy, mathematics, and surgery did not advance so far in this period that they freed natural inquiry from the strong mythological and magical roots that lay deep in man's cultural past. In fact, they but inched him toward a more naturalistic path of inquiry. When he undertook explanation of natural phenomena, he was inclined to fall back upon supernatural causation. Or when he undertook to explain the remote past, such as the origin and formation of the world, he had similar recourse to the supernatural. Thus his cosmogony, his cosmology, and his natural physics were intimately reflective of the supernatural. The magic, we have seen, entered immediately into his daily life in the use of incantations for therapeutics. We have also seen how astrological motives stimulated him to the scientific observation and description of the movements of heavenly bodies.

Chapter 2

Greek Science: Origins and Methods

MUCH HAS been written concerning the seemingly sudden emergence in the Greek Ionian colonies of the sixth century B.C. of a natural philosophy or science rational and surprisingly secular in character. In fact, historians, in recognition of the gulf that separates the approach of this natural philosophy from that of the cosmology of the Egyptians and the Babylonians, have called this phenomenon the "Greek Miracle." Study of the antecedents in earlier Greek and Near Eastern cultures does something to lessen the miraculous element but leaves us with considerable admiration for the first two centuries of Greek science and philosophy, from about 600 B.C. to about 400 B.C.

We cannot hope to detail here the developments leading to the "Greek Miracle." But we can suggest certain crucial factors. We recognize the importance of the change from a Bronze Age civilization to an Iron Age civilization, a change made possible by the improvement of the techniques for reducing and working iron. These improved techniques appear toward the end of the second millennium B.C. The cheap production of tools and weapons resulting from that change must have been an important factor in the ability of the smaller Greek city-states to compete successfully in trade with the more centralized monarchies of the Near East.

Of at least equal importance, no doubt, was the development of the alphabet, which tradition would have us believe originated in Phoenicia about 1200 B.C. and which spread to areas of Greek culture sometime after the turn of the millennium. Adoption of alphabetic writing did not, of course, initiate anything like popular education. But the comparative ease with which alphabetic writing can be learned certainly made possible a wider distribution of learning than had prevailed in the earlier monarchies, where writing and reading were the property of an exclusively scribal-priestly class.

The independence of the commercial-minded inhabitants of Miletus, an Ionian city on the coast of Asia Minor, the possible weakness of the ties between the governing classes of Miletus and older religious orthodoxy of the cities of the mainland, the immediate contact with the more cultured peoples of Asia Minor, who had drawn their culture from Mesopotamia and Egypt, and the occasional direct contacts with the culture of Mesopotamia and Egypt themselves—all these factors and no doubt others helped to produce the natural philosophy associated with the "school" of Miletus and its traditional founder, Thales (*fl.* 575 B.C.).

But lest we ignore the intellectual past of Greece in favor of uncertain, although probable, social factors, we should not set aside the undoubted effect of the changing mythology on the evolution of natural philosophy and science among the Greeks. H. Diels, the greatest of the modern editors of the fragments remaining from the writers of the pre-Socratic period, conjectures as to the very probable development of Greek thought out of earlier mythological speculations. And one of our most distinguished students of Greek culture, W. Jaeger, has led us carefully along the road from the mythological cosmogonies of early Greece to the "natural" theology that is evident among the so-called "materialists" of Ionia.

It is convenient to divide the period of Greek science into four main chronological divisions. The first and formative period is that usually called by the historians of philosophy the pre-Socratic period, from about 600 B.C. until just before 400 B.C. The second is the fourth century, the century of Plato and Aristotle and, later, of the creation of the Epicurean and Stoic philosophies. The third period is the so-called Hellenistic period, 300—100 B.C., when Greek culture behind the conquests of Alexander began to spread over the Near East and react more directly with the remains of the older cultures. This was the great period of Greek science, the period of Euclid, Archimedes, Apollonius, and many others. And the last is the Greco-Roman period, from about 100 B.C. to A.D. 600, a period in which Greek science was affected by the spiritual and nonrational currents that were in part responsible for the rise of Christianity. It was in this period that the Greek science which was to pass later to the Arabs and through them to the Latin West was epitomized, reorganized, and subjected to extensive commentaries. Hence this period must be studied in detail by the student of medieval and modern science. We have reserved the second part of this book to its study. In the first part of the book we shall analyze the important advances made during the first three periods, first as to method and then as to substantive knowledge.

Initially we must insist upon the general "rational," critical, often secular and nonmythological tone that the natural philosophers of the pre-Socratic period gave to much of Greek thought and science. We hasten to add that this does not mean that Greek philosophy in general was atheistic. The briefest reading of such Pythagorean fragments as appear genuine, the references in the Ionian fragments themselves, and the subsequent Neo-Platonic development make any such judgment ridiculous. But the critical spirit that emerges from this period is of great moment for the subsequent growth of science. Recalling the widespread attribution of disease among the Mesopotamians to demons or to displeased gods, one can feel that something profoundly significant has happened when the Hippocratic author of the *Sacred Disease,* writing toward 400 B.C. under the influence of the Ionian school, can open as follows:

> I am about to discuss the disease called "sacred." It is not, in my opinion, any more divine or more sacred than other diseases, but has a natural cause, and its supposed divine origin is due to men's inexperience, and to their wonder at its peculiar character. . . . (*Sacred Disease,* Chap. I, Loeb Classical Library translation of W. H. S. Jones, London, 1923.)[1]

Another distinctive feature of Greek thought that emerged during the first period was the basic concept of a "generalized" science as distinct from a set of empirical rules. It most clearly appears in the creation of a theoretical and abstract

[1] Quoted by permission of the present publisher and holder of the copyright, the Harvard University Press.

geometry. As we indicated in the last chapter, the Egyptians were accustomed in their mathematical papyri to give specific problems, such as the finding of the area of a particular field with particular dimensions. The theoretical geometry behind these empirical operations remained unexpressed and latent. Now with the Greeks it was the theoretical and abstract geometry that became the object of attention. They arrived at the general solution for the area of any triangle, starting with fundamental definitions, axioms, and postulates.

Although it is difficult to know to what extent these early Greek authors were aware of their invention of a generalized science, a much later author, Proclus, in speaking of the origins of geometry, says:

> It was Thales, who, after a visit to Egypt, first brought this study to Greece. Not only did he make numerous discoveries himself, but laid the foundations for many other discoveries on the part of his successors, attacking some problems with greater generality and others more empirically. . . . After these Pythagoras changed the study of geometry, giving it the form of a liberal discipline, seeking its first principles in ultimate ideas, and investigating its theorems abstractly and in a purely intellectual way. (M. Cohen and I. E. Drabkin, *A Source Book in Greek Science*, New York, 1948, p. 34.)[2]

Interestingly enough, the same author, in a passage just preceding the one quoted, had insisted upon the dependence of the Greeks on the Egyptian beginnings and also upon the birth of science in social needs:

> . . . it was, we say, among the Egyptians that geometry is generally held to have been discovered. It owed its discovery to the practice of land measurement. For the Egyptians had to perform such measurements because the overflow of the Nile would cause the boundary of each person's land to disappear. Furthermore, it should occasion

[2] This and the succeeding quotation are quoted by permission of the publisher, the McGraw-Hill Book Company, Inc.; copyright, 1948.

> no surprise that the discovery both of this science and of the other sciences proceeded from utility, since everything that is in the process of becoming advances from the imperfect to the perfect. The progress, then, from sense perception to reason and from reason to understanding is a natural one. (*Ibid.*, pp. 33-34.)

The debt of Greek geometry to the Egyptians had been asserted by authors much earlier than Proclus, the earliest being Herodotus. Nor was the nicely phrased judgment of Proclus on the precedence of utility to theory particularly original, since Aristotle had expressed the same idea almost one thousand years earlier.

Closely connected with the rise of the concept of a generalized, theoretical, and abstract science and closely connected also with the rising critical spirit among the pre-Socratic philosophers, particularly in the so-called Eleatic school of philosophy, was the evolution among the Greeks of a strict methodology of reason, or logic. Together with its kindred disciplines of mathematics, logic is a fundamental instrument of science. Observed data, whether assembled by the most careful experimental means or not, would mean little if we had no rules for testing the truth and falsity of arguments. It would, of course, be impossible to say when man first used rules of logic—say, for example, the principle of noncontradiction. But it is clear that conscious and critical study of the rules of reasoning is a Greek discovery. In the pre-Socratic period the founder of the Eleatic school, Parmenides (*ca.* 475 B.C.), would appear to be one of the first to be entirely self-conscious about the internal consistency of an argument. In fact, he insists upon it to the point that if the argument, when properly premised upon the inconceivability of the existence of nonbeing, leads us to conclusions that appear contrary to experience (one such conclusion being the nonexistence of movement), then we must ignore our experience and have confidence in reason:

> For this *(view)* can never predominate, that That Which Is Not exists. You must debar your thought from this way

of search, nor let ordinary experience in its variety force you along this way, *(namely, that of allowing)* the eye, sightless as it is, and the ear, full of sound, and the tongue, to rule; but *(you must)* judge by means of the Reason *(Logos)* the much-contested proof which is expounded by me. (Translation of K. Freeman, *Ancilla to the Pre-Socratic Philosophers,* Oxford, 1948, p. 43.)[3]

It is, then, among the Eleatic philosophers of the sixth and fifth centuries B.C. that we can find important beginnings in logic, particularly in deductive logic, which was used so skilfully by Plato, was formulated as a discipline by Aristotle, and served as the chief instrument for the extraordinary mathematics of Euclid and Archimedes—in fact, for the Hellenistic science generally (from about 300 B.C.). The Greeks, then, became masters of deduction, the drawing of *necessary* inferences from given premises.

On the other hand, less satisfactory was the Greek understanding of induction, the drawing of *probable* general conclusions from a multiplicity of particulars, the relative probability of the general conclusion depending on the relative completeness of the set of particulars. Also unsatisfactory was the discussion by the Greeks of the relation of arguments to experience, although in practice, as we shall see, they often exhibited an almost intuitive understanding of the proper relation of a scientific theory to observed data.

A word must be said about the very difficult question of Aristotle's scientific method. Although his *Posterior Analytics* is his chief discussion of this question, there is much elsewhere to throw light on it. In the first place the object of a science is to find principles, its elements, or its causes. This is as true of physics as it is of zoology. "The natural way of doing this is to start from the things which are more knowable and observable to us and proceed towards those things which are clearer and more knowable by nature." That is, we proceed initially from complex effects to simple causes; and once we have found causes or principles we have scientific knowl-

[3] Quoted by permission of the publisher, the Clarendon Press, Oxford.

edge. But how do we proceed to causes? Herein lie the difficulties of Aristotelian procedure.

Let us take first the case of finding out what we call "specific" causes or kinds. And among these we investigate initially the case where the cause we are seeking produces an action. Experience shows us that when an individual Callias was ill of some disease, a certain drug cured him. The same thing happened to Socrates and to many other individuals. As Aristotle says, this is a matter of experience. But then we do "science" when we conclude that such a drug generally cures all persons of a certain constitution who are afflicted by this disease. We have, in short, "from many notions, gained by experience, [produced] one universal judgment about a class of objects." A similar procedure is evident in forming conclusions about the causes of the characteristics of biological organisms. We are to proceed, Aristotle holds, after considering the phenomena presented by animals and their several parts, subsequently to treat of the causes and the reasons why. Although the individuals comprised within a species, such as Socrates and Coriscus within the species man, are the real existences, it is sufficient for the purpose of science to describe the universal attributes of the species, that is, the attributes common to all of its individuals. Let me give a final example of this initial procedure from individual to specific nature. From individual "lines" in nature we could induce the essence or cause of the species "straight line," namely, extension in either direction.

We can call this initial procedure of Aristotle "induction to species." But in science we do more than determine specific causes; we also determine more general causes—generic causes. Aristotle holds that if certain species comprise a genus and all the species are identifiable as a given cause, then that cause is generic. For example, let A stand for "long-lived," B stand for "bileless," and C stand for the particular kinds of long-lived animals, for example, man, horse, mule. A then belongs to the whole of C, for whatever is in C is long-lived. But B ("bileless") also belongs to all C. Hence A belongs to B, i.e., the long-lived animals are bileless. But demonstratively (rather than inductively) the cause they are long-lived is generic: they are bileless. But as Aristotle warns it must be apprehended that C is made up of all the particulars. "For induction proceeds through the enumeration of all the [specific] cases." This whole procedure is called the inductive syllogism.

We noted in the beginning of our statement in re Aristotle's method that in discovering scientific knowledge we proceed from things more knowable to us, i.e., from the complex, undifferentiated wholes given us as effects, to the things more knowable by nature, i.e., to the differentiated, elementary causes. But although Aristotle holds this procedure to be true for discovering scientific knowledge, in presenting the results of his scientific investigation, for example, in zoology, he starts with the most general causes, i.e., the attributes common to the largest groups of animals, which attributes he calls "analogues," to the next most general attributes which he labels "generic," and finally the least generalized attributes which he calls "specific" (see Chap. 4, Sect. II below).

We are not to think that the search for principles recommended by Aristotle is self-contained within each science, that once we have found the principles of the science we have necessarily reached the end of our investigation. In some sciences, as in optics, it is apparent that we have to go behind its principles or causes to another science, namely mathematics, for a further search for causes or principles.

Now a number of criticisms have often been leveled against Aristotle's methodology. They are essentially these: (1) The initial induction of a specific generalization which Aristotle lays to intuition constitutes a much more difficult and hazardous task than Aristotle seems to realize. (2) In his search for causes Aristotle often seems satisfied with finding merely names. In other words, his causes are often just restatements of the effects in terms of undefined powers producing these effects. (3) In physical science Aristotle remains too close to gross observation. In short, in physical science he leaves little room for abstraction. For this reason, he seldom applies mathematics to physical problems. Or to put it in another way, his physical science is too qualitative. (4) Once having found causes Aristotle does not have adequate methods of verification. He did not develop a technique which employs critical experiments to test other conclusions deducible from the principles. (5) And closely connected with this last, Aristotle seems to have little interest in prediction, the essence of modern scientific theories. For him science is the search for causes to explain the facts which are already at hand. This is what he intends when he calls science the syllogism of the reasoned fact.

A medical author of Aristotle's time, in the treatise entitled *Precepts,* while speaking in very general terms expresses the growing concern of physicians with the problem of the proper relationship of theory *(logismos)* and experience, a problem that occupied a number of philosophers and physicians from about 300 B.C. to about A.D. 200.

> . . . one must attend in medical practice not primarily to plausible theories, but to experience combined with reason. For a theory is a composite memory of things apprehended with sense-perception. For the sense-perception, coming first in experience and conveying to the intellect the things subjected to it, is clearly imaged, and the intellect, receiving these things many times, noting the occasion, the time and the manner, stores them up in itself and remembers. Now I approve of theorising also if it lays its foundation in incident, and deduces its conclusions in accordance with phenomena. (Hippocrates, *Precepts,* I, translation of W. H. S. Jones, London, 1923.[4])

We have already briefly suggested above that the Greeks often had a happy faculty for following what was essentially sound use of empirical observation as a foundation of and check on theory. For example, the whole course of Greek astronomy, as one mathematical system after another attempts to account for the solar, lunar, and planetary movements, was fashioned with the avowed intention of "saving the phenomena"—i.e., accounting for appearances—and it seems that one or another of the earlier systems was rejected

[4] Quoted by permission of the present publisher and holder of the copyright, the Harvard University Press.

precisely on the grounds that the theory could not account for the phenomena, or rather that the phenomena directly contradicted some deductive conclusion of the theory. Thus the system of concentric spheres introduced by Eudoxus in the fourth century B.C. and taken up and popularized by Aristotle foundered on the fact that the apparent size of the moon and sun, as well as the brightness of Venus and Mars, varies. This indicates that these bodies are not always the same distance from the earth, and they would have to be as a necessary conclusion of the theory of concentric spheres. One later astronomer, Sosigenes, categorically states this failure to account for phenomena as the reason for the decline of the theory of Eudoxus: "Nevertheless the theories of Eudoxus and his followers fail to save the phenomena. . . ."

In stressing the importance of observed data, carefully collected and utilized for scientific investigation, we should also note briefly Aristotle's magnificent zoological works and the amount of detail collected and used to support his attempts to generalize on the forms of living things. Although we shall mention these works again in talking about the substantial developments of Greek science, we can point out here, for example, the wealth of carefully noted detail brought to support his general classificatory statements, particularly where he discusses and generalizes on the various kinds of generation. Similarly rooted in careful observation were the botanical investigations of Aristotle's student, associate, and successor, Theophrastus, one of the most brilliant scientific figures in Greek antiquity.

Somewhat different was the role of observation and experience in the formation of the early systems of natural philosophy of Thales and his successors. These systems seem to have originated in the grossest analogies and patently insufficient observational data. Thus it is supposed that such gross facts as the plenitude of water on the earth's surface and its ready change of form to ice or vapor led Thales to assume water as the fundamental stuff of nature and to build a system around this assumption. It may be that the insufficiency of the ties between theory and experience in the early systems is merely representative of the fact that these were the first stages of science and philosophy. It was recognized as important that there be some ties between the theoretical explanation of nature and our experience of nature; but the necessity of a multiplicity of such ties and of their surety was not apparent. In a sense, the growth of modern science has been brought about on the one hand by the increasing sophistication of theoretical explanation, due largely to the use of mathematics, and on the other hand by the development of experimental ways to establish the surety and firmness of manifold bonds that unite theory with experience.

It would, of course, be incorrect to state that there was no experimentation in antiquity, either for the purpose of uncovering new facts about nature or for the purpose of confirming scientific theory. Even at the earliest stages of Greek science, in the sixth and fifth centuries B.C., there was experimentation by Pythagoras and the early Pythagoreans. Thus Pythagoras or his followers clearly established by experiment the relationship between the lengths of vibrating strings and the pitch of the notes emitted by the strings. It is true that the equally famous experiment of Empedocles (490-435 B.C.) with a water vessel to prove the corporality of air was more a notation of common experience than a deliberately planned and controlled test to confirm theory. But numerous controlled experiments are recorded in the Hippocratic medical treatises, which date from the fifth and fourth centuries B.C., and when we examine the activity of the successor of Theophrastus at the famous Lyceum, Strato the Physicist, we are confronted with activity deliberately experimental for purposes of scientific investigation. A later mechanical author, Hero of Alexandria, begins his discussion of a vacuum with a beautiful section which is thought to be drawn from Strato, and which has reference to numerous experiments:

> Vessels which seem to most men empty are not empty, as they suppose, but full of air. Now the air, as those who have treated of physics are agreed, is composed of particles, minute and light, and for the most part invisible. If, then, we pour water into an apparently empty vessel, air will leave the vessel proportioned in quantity to the water which

enters it. This may be seen from the following experiment. Let the vessel which seems to be empty be inverted, and, being carefully kept upright, pressed down into water; the water will not enter it even though it be entirely immersed: so that it is manifest that the air, being matter, and having itself filled all the space in the vessel, does not allow the water to enter. Now, if we bore the bottom of the vessel, the water will enter through the mouth, but the air will escape through the hole. Again, if, before perforating the bottom, we raise the vessel vertically, and turn it up, we shall find the inner surface of the vessel entirely free from moisture, exactly as it was before immersion. Hence it must be assumed that the air is matter. The air when set in motion becomes wind (for wind is nothing else but air in motion), and if, when the bottom of the vessel has been pierced and the water is entering, we place the hand over the hole, we shall feel the wind escaping from the vessel; and this is nothing else but the air which is being driven out by the water. It is not then to be supposed that there exists in nature a distinct and continuous vacuum, but that it is distributed in small measures through air and liquid and all other bodies. . . . The particles of the air are in contact with each other, yet they do not fit closely in every part, but void spaces are left between them, as in the sands on the sea shore. . . . Hence, when any force is applied to it, the air is compressed, and contrary to its nature, falls into the vacant spaces from the pressure exerted on its particles: but when force is withdrawn, the air returns again to its former position from the elasticity of its particles, as is the case with horn shavings and sponge, which, when compressed and set free again, return to the same position and exhibit the same bulk. Similarly, if from the application of force the particles of air be divided and a vacuum be produced larger than is natural, the particles unite again afterwards. . . . Thus, if a light vessel with a narrow mouth be taken and applied to the lips, and the air be sucked out and discharged, the vessel will be suspended from the lips, the vacuum drawing the flesh towards it that the exhausted space may be filled. It is manifest from this that there was a continuous vacuum in the vessel. . . . (Cohen and Drabkin, *A Source Book in Greek Science*, pp. 249-50.)[5]

The passage goes on in brilliant fashion to cite other experiments for confirming the existence of a vacuum spread between the particles of air and for producing a vacuum artificially. This tradition of experimental activity was carried on into the mechanical works of the Hellenistic period, where numerous experiments in pneumatics and mechanics were performed.

If, then, scientific investigation in antiquity involved considerable experimental activity, we may well ask why it was that Greek science falls short of modern science. It falls short in the maturity and the universality of its use of mathematical-experimental techniques. There is no question that a mathematical-experimental science existed in nascent form, at least, in optics, in statics, and in applied mechanics; that a mathematical-observational science was present in astronomy; and that an experimental science existed in zoology and physiology. But the techniques of these sciences were not yet commonly considered as the necessary methods in all fields of natural investigation. Before mathematical and experimental techniques had become the *common* property of Greek science, that science began to level off. The leveling off (note that reference is made to "leveling off" rather than to "dying out") of Greek science in late antiquity took place for a number of important political and social reasons, which will be discussed in detail in the next part of this volume. We can, however, note briefly the importance of Rome's rise to political power and domination of the Mediterranean area, the rise of Christianity and the consequent funneling off of many scholars who might have been scientists into dogmatic activities, and the general effect of noncritical spiritual forces that beset the Mediterranean world from at least late Hellenistic times.

In concluding our remarks on the general methodological achievements of Greek science we ought to mention Greek

[5] Quoted by permission of the publisher, the McGraw-Hill Book Company, Inc.; copyright, 1948.

contributions toward the establishment of institutions or organizations for scientific investigation. Some might say that the priestly-scribal classes of both Egypt and Mesopotamia preceded the Greeks in such organization. It has been conjectured on fair evidence that the best of the Babylonian cuneiform mathematical texts are products of scribal schools. And it would be interesting to know the full range of intellectual activity at the Royal Library of the Assyrian Assurbanipal.

But it appears that for Western science the Greek organizations and "schools" were of the greatest importance, because so many of the scientific works that were to form the basis of Western thought were produced there. Now it has long been a debatable question how formally organized were the Greek medical schools of the fifth century B.C., schools such as the great medical school at Cos. Some historians have held that these schools had formal organization of a guild nature. The existence of medical libraries connected with these schools has been argued persuasively. The Pythagoreans, at any rate, had formal organization, apparently of religious character, and if some of their activity was "scientific" much of their organized activity was religious. We can only guess as to the ties that held together the followers and successors of Thales at Miletus, or those of Democritus at Abdera, or other groups of philosophers in various places.

In tracing the growth of institutions of learning we cannot pass over Plato's Academy, which is, in one sense, the spiritual ancestor of Western academic organization, even as the name itself suggests. Founded after 388 B.C., it probably possessed little formal organization. From the tradition that there was erected in Plato's garden a shrine to the Muses has come the apparently ill-founded conclusion that the Academy was incorporated as a *thiasos,* or religious fraternity. It appears that regular dinners were held by the Platonic group, but probably few formal lectures. However, we simply do not know. Nor is the curriculum known, although efforts have been made to endow the early Academy with the educational systems described in the *Republic* and the *Laws.* Although we cannot be certain of the nature of formal instruction given in the sciences, we do know that a series of brilliant mathematicians and astronomers were associated with Plato. It is difficult to escape the conclusion that astronomical and mathematical activities were pursued vigorously at the Academy.

Even more clearly than the Academy of Plato was the Lyceum of Aristotle an institution for scientific research. It was founded after Aristotle's return to Athens in 335 B.C. We know a little of its organization in the time of Aristotle's student and successor, Theophrastus. It had a garden (perhaps for botanical or pharmacological research), a lecture room, an altar to the Muses. That lectures were given there can scarcely be doubted. The form of Aristotle's works is usually ascribed to their having been prepared as lecture notes. Subjects for investigation at the Lyceum were historical, political, literary, and scientific. Something more of the school's activities will be said in the discussion of Aristotle's and Theophrastus' contributions to biology (Chap. 4) and in the detailing of some of the experimental work and physical views of Strato (Chap. 6).

Both the Academy and the Lyceum had long if erratic lives that extended into late antiquity. The Academy was not permanently closed until A.D. 529. The general spirit of the two groups seems to be well reflected in the will of Theophrastus as he bequeaths property to the school:

> The garden and the walk and the houses adjoining the garden, all and sundry, I give and bequeath to such of our enrolled friends as may wish to study literature and philosophy there in common, since it is not possible for men to be always in residence, on condition that no one alienates the property or devotes it to his private use, but so that they hold it like a temple in joint possession, as is right and proper, on terms of familiarity and friendship. (W. Jaeger, *Aristotle*, 2nd ed., Oxford, 1948, p. 315.[6])

There is some historical evidence that the activity at the Lyceum led directly to the organization of the Museum at Alexandria, scarcely more than a generation after Alexander

[6] Quoted by permission of the publisher, the Clarendon Press, Oxford.

had swept over the Near East. In all probability the Museum began to take shape under the first king of the newly established Macedonian dynasty, Ptolemy Soter (305-283 B.C.), and a foundation date of about 290 B.C. would not be too wide of the mark. Unlike its predecessors, the Museum was a state-sponsored institution with a semireligious administrative organization dedicated to the Muses. Needless to say, it was not an organization for religious worship, in spite of the fact that its director is sometimes designated as Archpriest of the Museum. More than any of its predecessors the Museum was a center of scientific research. During the reign of the second Ptolemy, Ptolemy Philadelphus, its members numbered perhaps one hundred. Connected with the Museum was the great Alexandrian library, long the largest and most famous library in the world. It was built up in part at least by the purchase of private libraries, one of which, a doubtful tradition tells us, was that of Aristotle. The figures we have on the library's size are somewhat ambiguous but seem to show that under Ptolemy Philadelphus it had close to 500,000 books. We know next to nothing about the methods of instruction at the Museum. Possibly it had lectures, like the Lyceum. We know that colloquies or group discussions were held. It is tempting to think that there were laboratories of some kind, where the numerous experiments recorded of the Alexandrians were performed. Specimens of animals and plants seem to have been collected, but it is doubtful that there were public exhibits like those of the "museums" of a later day. The closest parallel of the Museum today would probably be a state research laboratory or a graduate school.

Chapter 3

Science and Early Natural Philosophy

To THIS point we have discussed without much attention to chronology some of the important methodological aspects of Greek science; we should now attempt to say something of the content of Greek science, discussing particularly those works which were to pass on to the Middle Ages and early modern times. Turning to the early pre-Socratic period, we ought to note initially that the first philosophers of Miletus set the basic problem for the philosophy of the succeeding generations—namely, the determination of the nature of the basic stuff *(physis)* of the universe. In the course of the search for the answer to that problem at least three characteristic ways of looking at nature emerged by the time of Aristotle; each has had enormous influence in the history of Western thought.

The first of these ways of looking at nature is called the physical or sometimes the material view. It emphasized the reality and permanence of matter and movement and, in its most mature form, the existence of void or empty space. Responsible for initiating this point of view were the Ionian philosophers of Miletus: Thales (*fl.* 575 B.C.) with his view of the fundamental character of water; Anaximander (*fl.* 547 B.C.), who subtly fell back on a fundamental "indeterminate" substratum out of which all things were formed and into which they returned unceasingly; and Anaximenes (*fl.* sixth and fifth centuries B.C.), who relied upon mist or air as the cornerstone of matter. In general the physical view held that the substance of knowledge was supplied by the sense, even though such philosophers as Heraclitus, who seemed to express general sympathy with the point of view, cautioned that "the eyes and ears are bad witnesses if the mind cannot interpret what they say," and the atomists built up a system that is not directly sensible but that is entirely inferential from sensory data. This view of nature reached its highest expression in the atomic theory of Leucippus and Democritus (fifth century B.C.). It sought to overcome the criticism of the Eleatic school (noted in Chap. 2), which denied movement as illusory, by positing the real existence of a void in which an infinite number of infravisible and indivisible atoms everlastingly moved, associating themselves now in one configuration, now in another. In spite of its general eclipse in the Middle Ages, this view of nature has never wanted some adherents. The reader will hardly have to be reminded of its success in modern times.

The second solution to the problem of nature has been given various names, which range from the psychical view of nature, or formal view, to the mathematical view. It emphasized the reality and permanence of forms, ideas, and concepts. The world of the senses, matter, was conceived to have a transitory nature. In the least subtle form of this view, the way to knowledge was to free oneself from the impermanence of the senses and by recollection to recover the knowledge of the permanent forms and ideas that reside in the soul. This view attained its earliest systematic expression with Plato. Some would say that Pythagoras (*fl.* 532 B.C.) and the Pythagoreans adumbrate the ideas implicit in this view of nature. But the Pythagorean fragments are ambiguous, and John Burnet, in emphasizing their experimental character and describing the general cosmology found in them, ties the Pythagoreans with the Ionians. The interest of the Pythagoreans in mathematics and their resorting to a belief which almost seems to make number itself the basic stuff of the universe, as well as their view that nature is in a sense written in mathematics, do provide background material for the Platonic view of nature. The consequences of this Pythagorean belief in the mathematical character of the universe have been enormous. Some twenty-two hundred years later Galileo was still to believe in it.

The third view of nature, that of Aristotle, can be characterized as a compromise view. Emphasis was laid on "becoming," on the actualizing of things that exist in potentiality. Nature is motion conceived in its broadest aspect as any kind of change. Rejecting the exclusive emphasis that the materialists put on matter and the Platonists on form, this view pictured matter and form as inseparable and inextricably bound together, and thus as equally important for the understanding of the fundamental character of nature. Aristotle seems to have been led to this view by the necessity of accounting for generation (coming into being) and corruption (passing away), and also for the kind of organization that we find in living organisms. We shall have much to say of the details of Aristotle's physics and something to say of his biology in succeeding chapters. For the Aristotelian framework became of paramount importance for medieval philosophy and science.

In the course of the evolution of these various views of nature, the pre-Socratics and their immediate successors in the fourth century B.C. advanced, often in very primitive form, some stock ideas that were to reappear again and again in the history of Western science and thought in constantly more subtle and expressive forms:

(1) The idea of the nondestructibility of the fundamental material substratum as it passes from one form to another lies behind the whole discussion of the basic stuff of the universe by the Ionians and their successors. When applied to chemistry, as it was by Jean Rey in the seventeenth century and Lavoisier in the eighteenth, this idea leaves the philosophical realm and becomes the fecund scientific principle of the conservation of the *weight* of matter passing from one chemical form to another. Although sharpened by the concept that all matter has weight, this principle is fundamentally tied with the broader philosophical idea.

(2) From the empirical observations of change of phase—say, from water to ice or to steam—and from the concept of the permanence of the substratum came the notion that change of phase brings with it changes in properties of basic matter. This idea, already present in fragments of Anaximenes and Heraclitus, has a long history in the development of alchemy and chemistry.

(3) Complementary to this idea was the thesis that change of phase and hence change of properties are accomplished by rarefaction and condensation. This appears to have been the basic idea behind Anaximenes' account of nature. It is

further employed in the atomic philosophy and has scientific significance in modern times.

(4) Constant attention to change and movement and speculation as to the causation involved stimulated the basic idea that when things change or move there are activating forces —no longer mythological but physical forces. The concept of force begins to get historical attention in the fifth century B.C., with Empedocles' (490-435) assignment of two fundamental and opposing forces, poetically named Love and Strife, and with the introduction by Anaxagoras, his contemporary, of *nous,* or mind, as an external force causing motion. The early emphasis on force, power, action is also reflected in the medical works contemporary with the philosophical activity. Thus Plato says that Hippocrates (b. at Cos in 460? B.C.) holds that to find the fundamental nature (*physis*) of a thing we must examine its "power," or *dynamis*—i.e., its capacity of acting or being acted upon—and in truth this idea is suggested in several places in the Hippocratic corpus. And later, with Aristotle, active and resistive powers or forces play an important role, as they do with all of the subsequent scientists in antiquity and the Middle Ages who are interested in mechanics.

(5) Another general idea that comes out of this early Greek philosophical speculation and that was to play an important role in early mechanics is that of an infinitude of space. When this space takes on the basic Euclidian three-dimensional character, it becomes the idealized or abstract space in which, according to Newton, bodies tend to move uniformly in a straight line when undisturbed by external forces. Infinite void or space is one of the fundamental tenets of the atomic system of the fifth century B.C. It reappears often among the commentators in late antiquity and the Islamic period and occasionally among commentators in the Latin West in the Middle Ages.

(6) With an infinity of space or void and an infinity of matter in that space, we see the idea develop of an infinitude of worlds like our own. The discussion of a plurality or infinity of worlds, either successively or simultaneously existing, goes back to the atomists and perhaps even further; it appears again

and again before it comes to a head in the sixteenth-century treatment by Giordano Bruno.

Turning from the more general cosmological stock ideas evolving out of this early speculation, we should note at least two important fundamental ideas regarding living things that were to be enormously influential in the growth of science.

(7) The first is a speculative "hunch" as to the evolutionary nature of the development of living things, a hunch that is first assigned to Anaximander, that is rejected completely by Aristotle, but that reappears strongly much later, in a passage from Diodorus Siculus.

(8) The second is one of the basic biological ideas apparent in the Hippocratic corpus of medicine of the fifth and fourth centuries B.C.—namely, the concept of the healing power of nature. The physician is but to aid nature in effecting the restoration of the patient to health. This is the primitive recognition of the fundamental biological doctrine that organisms have the capacity or at least the tendency when damaged or diseased to return to their normal integrity by knitting broken bones, replacing tissue, etc.

In terminating these remarks on some of the stock ideas that appear to come out of earlier philosophical writers on nature, it is advisable to leave a word of caution. Our knowledge of the philosophy and science of the pre-Socratic period is based on a very scattered and meager set of fragments, together with the evaluations made of the opinions of the early philosophers by men of later periods who speak in the terms and disputes of the science and philosophy of their own day. And it is a favorite academic sport to search the fragments for clues of the origin of a favorite point of view. But with these facts considered, it is safe to say that we can find these germinal ideas in the early fragments or in the works of men not very far removed in time from the early philosophers.

Chapter 4

Greek Medicine and Biology

I

LEAVING the general philosophical ideas of the early periods of Greek thought that seem to have had later scientific importance, we can begin to describe the state and development of the individual sciences from about 400 B.C. to about A.D. 200, thus embracing all of the periods we have named as the second and third in our chronological outline and part of the fourth. We can point first to the development of medicine. We have already intimated that the Greeks created a secular medicine that often employed the methods of observation and experience. The tradition of rational and even experimental medicine (although there is some doubt of this) is usually extended back to the beginning of the fifth century B.C., to the natural philosopher and physician Alcmaeon of Croton, in southern Italy, perhaps a Pythagorean.

> His detailed work on the nature of the senses was based on experiment; tradition records that he was the first to "dare to undertake the excision of an eye." This clearly refers to a surgical operation on a human subject; and he profited by this opportunity, and doubtless also by dissection, to examine the eye and the brain. (K. Freeman, *The Pre-Socratic Philosophers,* Oxford, 1946, p. 137.[1])

A number of medical schools seem to have grown up in the course of the fifth century, of which the one that achieved the greatest fame both in its own day and with posterity was the school on the island of Cos. The works that emerged from this school were codified quite early under the name of the greatest physician of the school, Hippocrates, who flourished about 420 B.C. The so-called Hippocratic corpus contains some seventy works, which date from the time of Hippocrates to about 300

[1] Quoted by permission of the publisher, the Clarendon Press, Oxford.

B.C. The problem of deciding which of the works, if any, are genuinely by Hippocrates is generally considered insoluble on the basis of the now known evidence. The corpus as a whole was of extraordinary influence during late antiquity and the Middle Ages, both among the Arabs and in the West. Besides including the remarks quoted above on the importance of experience as a basis for medicine and on the secular character of epilepsy, the corpus contains a number of other ideas interesting to the historian of science. One treatise, in listing foods for dietary purposes, sets forth a very rudimentary classification of plants and animals. We find in another treatise, *On Nutriment,* of about 400 B.C., one of the earliest references in medical literature to the pulse as a diagnostic sign, as well as the beginnings of the distinction made in antiquity between the arteries and the veins. In certain of the treatises we find the development of the doctrine of humors, or body fluids, that later became so widespread. The humoral doctrine is epitomized in the work *On the Nature of Man*:

> The body of man has in itself blood, phlegm, yellow bile, and black bile; these make up the nature of his body, and through these he feels pain or enjoys health. Now he enjoys the most perfect health when these elements are duly proportioned to one another in respect of compounding, power and bulk, and when they are perfectly mingled, Pain is felt when one of these elements is in defect or excess. . . . (Loeb Classical Library translation of W. H. S. Jones, *Hippocrates,* Vol. IV, p. 11.[2])

With a pathology based on excess or defect of body fluids or some similar disharmony, Hippocratic therapeutics concerned itself with the restoration of the normal proportion or mixture of fluids or the general harmony of the body. And even though nature tended itself to restore the balance, the physician actively aided by the use of purgatives, fomentations and baths, barley waters, and wine, and by bleeding or venesection. Perhaps the greatest achievement of Hippocratic medicine lay in

[2] Quoted by permission of the present publisher and holder of the copyright, the Harvard University Press.

the masterful clinical descriptions that were to help the physician predict the course of disease. Who cannot feel the terseness and accuracy of the so-called *facies Hippocratica*, the signs of approaching death: "nose sharp, eyes hollow, temples sunken, ears cold and contracted with lobes turned outward, skin tense and parched, face discolored, eyelids livid, mouth open, lips loose and blanched."

With the spread of Greek culture and the rise of science in Alexandria, some remarkably original work was done in medicine that has a distinct bearing on the history of biology. Special attention should be paid to two important figures of the third century B.C. The first is Herophilus, who made anatomical descriptions that were in all probability a considerable advance over previous ones. That he was able to dissect human cadavers seems likely. At any rate, we are told that he left detailed descriptions of the various systems—nervous, vascular, digestive, and osseous. Particularly noteworthy was his description of the brain, in which the cerebrum was distinguished from the cerebellum. Herophilus discovered and named the duodenum, the first part of the small intestines; he distinguished the arteries from the veins, but showed that the former do have some amount of blood in them, contrary to some contemporary opinion which classified the arterial system as a carrier of a pneuma, or spirit, alone. His junior contemporary, Erasistratus, continued the work of Herophilus in anatomy and physiology. He described correctly the action of the epiglottis in preventing entrance of food and drink into the windpipe during swallowing. It is said that he rejected all but mechanical explanations in describing physiological functions. Thus he thought of food as being ground up in the belly during digestion. Even his explanation of the blood system, which was a distinct forerunner of the influential explanation of Galen, was formulated in mechanical terms, and this in spite of his use of airs, or pneumas. It appears that for him blood was the source of matter in the body, whereas pneuma (air as modified in the parts of the body) was the source of energy or activity. He explains why, although the arteries carry mostly pneuma, we find so much blood in them when they are cut. According to him, the blood gushes over into the arterial system when a temporary vacuum is formed there by the escape of the pneuma through the cut. In this he seems to be leaning on the physical ideas of Strato the Physicist, who apparently had contact both with the Lyceum at Athens, as a successor of Theophrastus, and with Alexandria, as a tutor of Ptolemy Philadelphus, second of the Greek rulers of Egypt. (See Chap. 6.)

We have the names of a number of the followers of these early Alexandrians and of the sect that arose. But we can make only passing reference to two of the chief sects which differed in their views toward method—that is, toward the relation of experience and theory in medical practice. The first sect was the Dogmatists, who, as Celsus (see Chap. 8) tells us, professed "a reasoned theory of medicine." This group "propounds as requisites, first, a knowledge of hidden causes involving diseases, [and] next of evident causes." Naturally they admit the evidence of experience—for example, in discussing the disease of internal parts where it "becomes necessary to lay open the bodies of the dead and to scrutinize their viscera and intestines. They hold that Herophilus and Erasistratus did this in the best way by far, when they laid open men whilst alive—criminals received out of prison from the kings —and whilst these were still breathing, observed parts which beforehand nature had concealed, their position, colour, shape, size, arrangement, hardness, softness, smoothness, relation, processes and depressions of each, and whether any part is inserted into or is received into another."

The second important sect that was established at Alexandria was that of the *Empirici*, or Empiricists. They accept evident causes, but, according to Celsus, "they contend that inquiry about obscure causes and natural actions is superfluous, because nature is not to be comprehended." They object to simple universal causes and reject entirely theoretical medicine, for "even philosophers would have become the greatest of medical practitioners, if reasoning from theory could have made them so."

Finally, we can move all the way to the second century after Christ for our concluding remarks on Greek medicine and its relation to the history of science. In doing so, we skip a number of interesting writers to take up the most important single figure in Greek medicine after Hippocrates—namely, Galen. Galen was not really a member of any medical sect, but he was probably closer in spirit to the Dogmatists than to the Empiricists. This, of course, did not prevent him from being one of the greatest medical experimenters in the history of early medicine. We can illustrate this fact by quoting from his demonstration of the irreversibility of the flow of urine from kidney to bladder:

> Now the method of demonstration is as follows. One has to divide the peritoneum in front of the ureters, then secure these with ligatures, and next, having bandaged up the animal, let him go (for he will not continue to urinate). After this one loosens the external bandages and shows the bladder empty and the ureters quite full and distended—in fact almost on the point of rupturing; on removing the ligature from them, one then plainly sees the bladder becoming filled with urine.
>
> When this has been made quite clear, then, before the animal urinates, one has to tie a ligature round his penis and then to squeeze the bladder all over; still nothing goes back through the ureters to the kidneys. Here, then, it becomes obvious that not only in a dead animal, but in one which is still living, the ureters are prevented from receiving back the urine from the bladder. These observations having been made, one now loosens the ligature from the animal's penis and allows him to urinate, then again ligatures one of the ureters and leaves the other to discharge into the bladder. Allowing, then, some time to elapse, one now demonstrates that the ureter which was ligatured is obviously full and distended on the side next to the kidney, while the other one—that from which the ligature had been taken—is itself flaccid, but has filled the bladder with urine. Then, again, one must divide the full ureter, and demonstrate how the urine spurts out of it, like blood in the operation of venesection; and after this one cuts through the other also, and, both being thus divided, one bandages up the animal externally. Then when enough time seems to have elapsed, one takes off the bandages; the bladder will now be found empty,

and the whole region between the intestines and the peritoneum full of urine, as if the animal were suffering from dropsy. Now, if anyone will but test this for himself on an animal, I think he will strongly condemn the rashness of Asclepiades, and if he also learns the reason why nothing regurgitates from the bladder into the ureters, I think he will be persuaded by this also of the forethought and art shown by Nature in relation to animals. (Galen, *On the Natural Faculties,* I, 13, Loeb Classical Library translation of A. J. Brock, London, 1916.[3])

Galen was enormously prolific as a writer, and his works deal with a number of philosophical questions not connected with medicine. He may have dissected human cadavers; but even if so, he dissected animals much more frequently, including Barbary apes and pigs. Many of his mistakes in anatomy and physiology derive from the fact that the analogies were too close that he attempted to draw between man and the animals he dissected.

One of the most interesting and influential of Galen's descriptions is that of the great blood and pneuma systems that had been in the process of formation since at least the fourth century B.C. As the reader well knows, the current theory of the circulation of the blood, which dates from the time of Harvey and Malpighi, in the seventeenth century, holds that the blood flows, or circulates, in an essentially closed system of arteries and veins, with the heart acting as a pump that motivates the flow. The bright scarlet blood charged with oxygen is pumped out of the left side of the heart through the arteries and then is carried throughout the body in the arterial system. From the arteries it passes through numerous systems of small capillaries, or tubes, in the tissues of the body into the veins. In the course of this passage the blood supplies oxygen to the tissues and changes to the darker hue that it has in the veins. In the veins it flows to the right side of the heart. From the right side of the heart it is pumped through the pulmonary artery to the lungs, where it is replenished with oxygen

[3] Quoted by permission of the present publisher and holder of the copyright, the Harvard University Press.

as it is forced through capillaries into the pulmonary vein. From the pulmonary vein it enters the left side of the heart, completing the circulation. Now let us return to Galen's description of the blood.

Like Erasistratus, Galen holds that blood and pneuma are

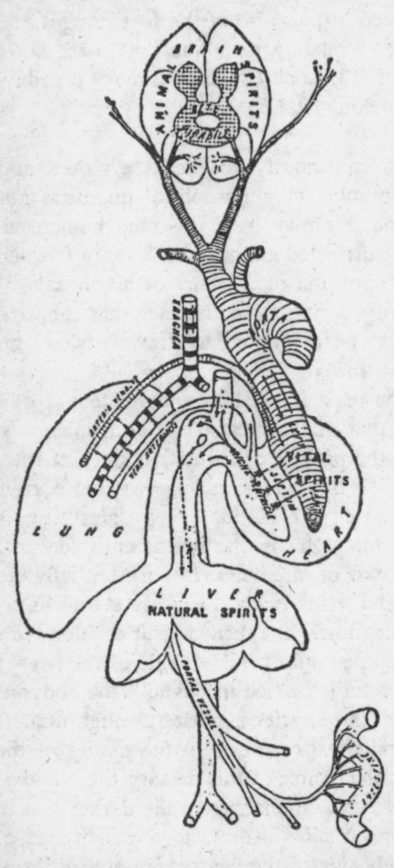

FIG. 7. Charles Singer's representation of Galen's vascular systems. (Reproduced from R. W. Livingstone, ed., *The Legacy of Greece*, Oxford, 1921, by permission of the publisher, the Clarendon Press, Oxford.)

the body's essential ingredients. The blood is fashioned in the liver from the chyle produced in the intestines (see Fig. 7). It is charged with at least traces of pneuma, which have their ultimate origin in the air we breathe. The blood is distributed to all parts of the body by the venous system in the main, but perhaps also in part by the arterial system. It is so carried to the right ventricle of the heart. Galen believed a certain amount of the blood trickles through pores or passages in the septum separating the right and left ventricles of the heart, and also that there were intervascular connections between the venous and arterial systems that allowed for the passage of more volatile elements. Some later Galenists would hold that this blood is charged with natural spirits or pneuma (although Galen does not make definite the role of natural spirits). Then the arterial system carried a mixture of blood and zotic or vital spirits throughout the body; the left ventricle of the heart together with the arteries constituted the seat or home of the innate heat of the body. Some of the blood and vital spirits reached the brain, where a psychic pneuma is formed. This psychic pneuma is then carried throughout the body by what was believed to be a third vascular or tubular system, the nervous system. A recent author has described some of the details of Galen's view of the blood system as follows:

> The veins serve the main purpose of anadosis, i.e. transportation of proper nourishment to the organs. The chyle from the gastrointestinal tract reaches the veins of the portal system where it appears as crude blood. The main process of hematopoiesis (blood formation) is effected in the liver, the supposed origin of the veins, which now offer to each part what it will attract, hold, assimilate, and eliminate by means of the natural faculties. In addition, the veins contain pneuma from three main sources. In the first place, the veins take the vaporous and useful part of the winds developed in the abdomen. In the second place, the juices of chyle and blood can and do exhale some pneuma, so that even the psychic pneuma, to a small extent, may be engendered from the veins in the cerebral ventricles. This is of importance in as far as the pneumatic content of the venous blood can sub-

stitute for the vital pneuma of the arteries. At any rate, the blood in the veins to some extent is vaporous; for its "thinnest and most vaporous" portion is attracted by the arteries through their anastomoses (intercommunication of blood vessels) with the veins. The arterial blood in turn, through these anastomoses, forms a third source of pneuma for the veins. "The arteries and veins form synastomoses in the whole body (i.e. where the walls of the two kinds of blood vessels border upon each other) and accept from each other blood and pneuma through some invisible and perfectly narrow paths." The role which the pores in the intraventricular septum of the heart are supposed to play in this respect is too well known to need any comment. It is only necessary to add that in Galen's opinion the arteries which reach the intestines on their part also absorb a little food in order to see the far-reaching parallel between arterial and venous blood. Neither of the two is without a pneumatic ingredient; the difference is one of degree. Whereas blood and pneuma are fine and thin in the arteries, they are crude and "smoky" in the veins. And whereas the arterial blood has a large pneumatic component, the latter is much smaller in the veins. Just because the difference is only one of degree, both arteries and veins can supply the organs with food. Generally speaking, the organs take thinner food from the arteries, thicker food from the veins. In particular, the nature of the organ will decide the kind of blood needed. The spongy spleen shows a wide ramification of large arteries and the lungs receive blood from the right ventricle of the heart where it has assumed an almost arterial character. The liver, on the other hand, has small arteries chiefly for cooling purposes. They do not absorb blood from the liver, nor do they have to feed its tissues with thin and vaporous food, "nor do they have to furnish the liver with much vital pneuma as they do some other organs." (O. Temkin, "On Galen's Pneumatology," *Gesnerus,* 8 [1951], pp. 184-85.)

Galen was of special importance for medieval science, for it is his presentation of Greek medicine that, with the Hippocratic corpus, was to be translated and studied in both Islam and the West. And it was his emphasis on the teleological or purposeful function of all the parts and organs of the body that was to appeal so greatly to both Islamic and Christian physicians.

II

The beginnings of the study of human biology, we have seen, appear in the medical works of antiquity. Those parts of biology which are designated as natural history, particularly zoology and botany, emerge as separate sciences in the fourth century B.C. Chiefly responsible was Aristotle (384-322 B.C.). Son of a physician, the great philosopher of Stagira, a town in northeastern Chaldice, came to Athens to study with Plato at the age of seventeen and stayed with Plato until the latter's death (347 B.C.). The impression made upon Aristotle by Plato was profound, and the dialogues written by Aristotle in this period and now lost were Platonic in spirit and content, according to our scanty knowledge of them. Leaving Athens, Aristotle studied and probably taught successively at Assos and Lesbos, in Asia Minor, then proceeded to the court of Philip of Macedonia to act as a tutor to the young Alexander, the future conqueror of Greece and the whole Near East. This period of wandering terminated with Aristotle's return to Athens in 335 B.C. On his return he founded the famous Lyceum, which became the first great scientific research center in the history of mankind, with the possible exception of Plato's Academy. But the latter appears to have been a place of philosophical inquiry—which included, of course, mathematics—more than it was a school for the patient collecting and organizing of scientific data, as the Lyceum clearly became.

Jaeger, one of the greatest students of Aristotle, has advanced the persuasive idea that in the course of the latter's career his approach developed from one that was predominantly Platonic and metaphysical to one that was increasingly empirical and more interested in what the Greeks call *historia* —i.e., research. Thus Jaeger believes that it was during the Lyceum period that Aristotle pursued further the research into and organization of zoological data which he probably had begun with Theophrastus, his associate and successor, on the island of Lesbos in his middle period. Presumably Aristotle's great *Parts of Animals* and *History of Animals* took final form at the Lyceum. The botanical works of Theophrastus were probably begun during the middle and last periods of Aristotle's life. To the period of the Lyceum belong, in all probability, Aristotle's short biological tracts, called the *Parva naturalia,* his *Meteorology,* and a lost work on the flooding of the Nile, of which we have just one brilliant fragment that disposes of the problem much mooted by Greek natural philosophers: "The Nile floods are no longer a problem, for it has actually been observed that rains are the causes of the swellings."

Patient work in chronological studies was also undertaken at the Lyceum, such as collection of the names of the winners of the Pythian Games and assemblage of the records of dramatic performances at Athens. *Historia* at the Lyceum is no doubt illustrated also by the collection of 158 constitutions of the various city-states, of which the extant *Constitution of Athens* forms one book. It was also at the Lyceum that first researches into the history of science were conducted, Theophrastus undertaking the history (partly extant) of the opinions of the *physici,* or natural philosophers, Menon describing the history of medicine (only extracts extant), and a third student, Eudemus, dealing with the history of arithmetic, geometry, and astronomy (only extracts extant). The schematism or systematization undertaken at the Lyceum was of great importance; but, according to Jaeger,

> . . . to suppose its real achievement lay in the classification of animals and plants would be wrong. It was far less important in the development of natural science than the fact that here for the first time the observation and description of the individual and its life history was being taken absolutely seriously. (W. Jaeger, *Aristotle,* 2nd ed., Oxford, 1948, p. 331.[4])

But what were some of the important biological results of

[4] Quoted by permission of the publisher, the Clarendon Press, Oxford.

Aristotle's research? The most important of his works containing the results of his researches into natural history are those already mentioned, the *History* [or *Investigation*] *of Animals* and the *Parts of Animals,* together with *On the Generation of Animals* and *On the Progressive Motion of Animals.* His more philosophical *On the Soul* is also important for understanding his views on life.

We are not to think that, because he paid more attention to detailed research toward the end of his career, Aristotle abandoned some of the important philosophical principles that had no doubt developed by the time of his middle period. In a sense the earlier periods gave him a certain fundamental metaphysical framework which he amended as the result of his research, but did not necessarily abandon. His ideas of matter and form, of the various kinds of causation, of movement as the actualizing of a potentiality—all play an important role in his biological researches. But more than once he stresses the importance of appealing to facts in sufficient number before forming general conclusions. Thus in the *Generation of Animals* in taking up a difficult problem he observes: "There are not enough facts to warrant a conclusion, and more dependence must be placed on facts than on reasonings, which must agree with facts." In another place he points out that it is easy to distinguish those who argue from facts and those who argue from notions.

Aristotle's collecting of the facts in natural history was done in a number of ways. He made countless personal observations of the habits and structures of animals. Some of the facts observed could have been made plain only by the use of the dissecting knife—e.g., his description of the eye of the mole, his description of the gall bladder of the pelamid, his description of the stomach in ruminants, or cud chewers.

In certain cases there are evidences of the dissection of living animals—i.e., vivisection. Thus there is reference to the movement of the heart of a tortoise after its removal and another to the movements of the heart and the sides of a chameleon after dissection. There is also much evidence that Aristotle used (often with critical caution) the observations and reports of others.

Some of the specific observations that Aristotle made have long stimulated the admiration of naturalists. An example is his distinction of cetaceans (whales, dolphins, and porpoises) from fish. He noted that the former have lungs and blowholes, that the embryo is tied to the mother by an umbilical cord, and that the young are born alive. His description of the placental dogfish was not confirmed until the nineteenth century, nor was his description of the sexual reproduction of the cuttlefish and the octopus, with the special arm acting as a sexual organ and fitting into the mantle-cavity of the female. This list of important observations can be extended easily: Aristotle described carefully the habits of bees, the development of the embryo of the chick, the transformation of the caterpillar to the butterfly, the testes of mammals, the habits of electric fish, etc.

But Aristotle's zoological works are more than mere description of animal habits. Certain fundamental ideas emerge from his descriptive material. Thus we see him supporting with some elegance the doctrine of epigenesis, which holds that the embryo develops from undifferentiated material. Again, in the *History of Animals* Aristotle is constantly concerned with the problem of the samenesses and differences of animals, the problem of classification; and, although he nowhere presents a single classificatory system, he makes numerous important statements on classification that allow us to construct a general system. Although we cannot deal with this system in detail, we should note that he recognizes the most general division possible between red-blooded (sanguinous) and non-red-blooded animals. This division corresponds to the modern distinction between vertebrates and invertebrates, since, as Aristotle had observed, it is the sanguinous animals that have a backbone. His classificatory statements involve the use of the terms *eidos* (species), *genos* (genus), and *analogon*. The first corresponds fairly closely to the use of the term "species" in modern times; the last two cut across classes, orders, and families, and are used loosely.

We have, then, first to describe the common functions, common, that is, to the whole animal kingdom, or to certain large groups, or to the members of a species. In other words,

> we have to describe the attributes common to all animals, or to assemblages, like the class of Birds, of closely allied groups differentiated by gradation, or to groups like Man not differentiated into subordinate groups. In the first case the common attributes may be called analogous, in the second generic, in the third specific. (Aristotle, *On the Parts of Animals,* I, 5, translation of William Ogle, Oxford, 1912.[5])

Aristotle's general ideas on classification had great vitality until the eighteenth and early nineteenth centuries, long after his authority in physics and philosophy was overthrown. His concern with classification seems to have led him to the recognition that nature can be ordered into a hierarchy of living things, a so-called ladder of nature from the simple to the complex:

> Nature proceeds little by little from things lifeless to animal life in such a way that it is impossible to determine the exact line of demarcation, nor on which side thereof an intermediate form should lie. Thus, next after lifeless things in the upward scale comes the plant, and of plants one will differ from another as to its amount of apparent vitality. . . . Indeed, as we have just remarked, there is observed in plants a continuous scale of ascent towards the animal. So, in the sea, there are certain objects concerning which one would be at a loss to determine whether they be animal or vegetable. . . . Thus the pinna (a mollusk) is rooted to a particular spot, and the solen (or razor-shell) cannot survive withdrawal from its burrow. (Aristotle, *Historia animalium,* VIII, 1, translation of D. W. Thompson, Oxford, 1910.[5])

Notice that Aristotle in this passage speaks of the varying amounts of vitality. The over-all ordering of living things rests fundamentally on the amount of vitality. Thus plants are distinct from animals in having limited vital activity, and animals are generally distinguished from man in the same fashion, as they possess more restricted vital activity than man. Aristotle's

[5] This and the succeeding quotation are quoted by permission of the publisher, the Clarendon Press, Oxford.

discussion of vital activity, so important for his biology, is concentrated in his book *On the Soul*. The soul for Aristotle is the life principle. "It is the form of a natural body having in it the capacity of life." As the form it is inseparable from the matter of the natural body. But in investigating the soul we are investigating the functions, or levels of vital activity, of living organisms, and for Aristotle this is psychology.

> We take, then, as our starting-point for discussion that it is life which distinguishes the animate from the inanimate. But the term life is used in various senses: and, if life is present in but a single one of these senses, we speak of a thing as living. Thus there is intellect, sensation, motion from place to place and rest, the motion concerned with nutrition and further, decay and growth. (Aristotle, *On the Soul*, II, 8, translation of R. D. Hicks.[6])

Now a plant possesses life in the sense of growth, decay, and reproduction only; i.e., the faculty of the vegetable soul is that of nourishment and growth for the purpose of reproduction. All living things possess this faculty in common with the vegetable soul, but the vegetable soul possesses only this faculty. The animal soul ordinarily possesses, in addition to the faculty of growth and decay, the faculties of local motion and sensation, "but it is sensation primarily which constitutes the animal. For, provided they have sensation, even those creatures which are devoid of movement, and do not change their place, are called animals." Finally, the rational soul of man possesses, in addition to the faculties of growth and decay, motion and sensation, that of intellect. This analysis, based in part on prior views of Plato, was the starting point for much discussion in the Middle Ages, in which the vegetative, animal, and rational souls emerged much more clearly as distinct and separable entities.

A final but important note on Aristotle's biology should be recorded. Aristotle, like Galen, was a teleologist. He believed firmly in the purposeful element of nature. "Therefore action

[6] Quoted by permission of the publisher, the University Press, Cambridge, England.

for an end is present in things which come to be and are by nature." Arguing against chance growth as being fundamental in nature he says, ". . . natural things either invariably or normally come about in a given way; but of not one of the results of change and spontaneity is this true." In his zoological works Aristotle investigates not only the purpose of the organism as a whole but also the purpose or end of the parts.

III

The work of Theophrastus stands clearly as a continuation and often brilliant extension of that of Aristotle. Although we know Theophrastus chiefly for his magnificent botanical works, his excellence in logic has recently been demonstrated and the importance of his metaphysical judgments has long been recognized. In natural history his reputation rests upon two great works: *History [or Inquiry] of Plants* and *Causes of Plants,* both of which may be later compilations of a number of works of Theophrastus. These are the most important products of botanical investigation until well into early modern times.

Like Aristotle, Theophrastus concerns himself with questions of classification, for "our study becomes more illuminating if we distinguish different kinds." Plants he classifies into trees, shrubs, undershrubs, and herbs. He discusses the various possible bases of classification of plants—such as wild or cultivated, aquatic or terrestrial, fruit-bearing or fruitless, flowering or flowerless, evergreen or deciduous—criticizing each one. But he concludes that in spite of our inability to be precise about definition and classification we should use these distinctions, since they are based on natural character, even though the character may be modified in different localities under different conditions.

Theophrastus' minute descriptions of plants are excellent, and he did much to establish a standard botanical language, some of which is still with us today. His discussion of the various methods of plant reproduction is of interest. "A plant has the power of generation in all of its parts, for it has life in them all." He distinguishes reproduction from a seed, a root, a piece torn from a branch or twig or from the trunk itself. Like Aristotle, he believed that on occasion spontaneous generation is possible. But at the same time he is cautious about the matter, and thinks that in some instances where spontaneous generation is assumed there is a more plausible explanation. For example, in certain cases where plants appear to grow up spontaneously, the seed has been carried there by the wind. In discussing germination he makes the important distinction between monocotyledons and dicotyledons—i.e., between those plants the seed of which has a single leaf, such as cereals, and those the seed of which has two leaves, such as leguminous plants.

Theophrastus' influence on medieval botany was extensive, but entirely indirect. A pseudo-Aristotelian work entitled *On Plants,* probably dating from the first century B.C., shows Theophrastian influences and was quite widely read in the Middle Ages. The same can be said for the great *Materia medica* of Dioscorides, which attempts a description and a rational classification of some six hundred plants, but the main concern of which is the plant as a medicine. Most of the medieval herbals used Dioscorides as a point of departure. It appears that Theophrastus' works were not translated into Latin until the fifteenth century.

Chapter 5

Greek Mathematics

IF THE high point of natural history came during the century of Aristotle and Theophrastus, that of mathematics and the sciences dependent upon it, such as physics and astronomy, came during the Hellenistic period, from about 300 B.C. But much important mathematical activity preceded the Hellenistic geometry and made it possible. We have already spoken of the invention of a general and abstract mathematics during the sixth and fifth centuries B.C.; now we should note briefly some of the developments that led to the flowering of mathematics in the Hellenistic period. During the earlier period, from the time of Thales onward, many specific theorems were discovered that were later to reappear in Euclid. In the discovery and elaboration of the proofs of these theorems not only was much of the language of geometry devised but some of the fundamental methods of discovery and proof evolved.

One of the most important of the general methodological achievements of the early Greek mathematicians was the development of the methods of analysis and synthesis, for these methods constitute the basic inferential procedure of Greek geometry. Analysis, according to the Greeks, commences with the assumption of what is to be proved and then proceeds backward by successive inferences to theorems or axioms or postulates generally accepted or previously proved. Synthesis is, of course, the reversal of this procedure, starting with the previously accepted or proved theorem and proceeding therefrom to the proof of the new theorem. As formal methods applied to higher problems, analysis and synthesis (particularly where they are together with analysis preceding and demanding a consequential synthesis for the proof of a particular theorem) are usually attributed to Euclid and his successors. But it is obvious that these basic methods are the product of the earlier geometrical activity of the fifth and fourth centuries B.C.

The classic Greek description of combined analysis and synthesis is that of Pappus of Alexandria (*fl.* at the end of the third and the beginning of the fourth centuries):

> *Analysis,* then, takes that which is sought as if it were admitted and passes from it through its successive consequences to something which is admitted as the result of synthesis: for in analysis we assume that which is sought as if it were (already) done and we inquire what it is from which this results, and again what is the antecedent cause of the latter, and so on, until by so retracing our steps we come upon something already known or belonging to the class of first principles, and such a method we call analysis as being solution backwards.
>
> But in *synthesis,* reversing the process, we take as already done that which was last arrived at in the analysis and, by arranging in their natural order as consequences what were before antecedents, and successively connecting them one with another, we arrive finally at the construction of what was sought; and this we call synthesis. (T. L. Heath, *Euclid: The Elements,* I, p. 138.[1])

One form of analysis—considered generally—that was invented by the Greeks and used with particular skill by Euclid and Archimedes was proof by reduction to absurdity. We assume the contradiction of what is to be proved and then by a series of successive logical inferences finally reach some inference that contradicts a theorem or first principle we hold to be true. Then, since we have proved that the contradiction to what we are proving is false, the theorem under proof is assumed to be true. The correct use of this type of proof corresponds with the very highest point of Greek geometry. It is also interesting to note that some of the very best geometrical proofs that are original with medieval authors are cast in this form, which, of course, the medieval geometer borrowed from Hellenistic geometry. (Consult Appendix I for an example of Archimedes' skillful use of this method.)

[1] Quoted by permission of the publisher, the University Press, Cambridge, England.

As we shall note later, in discussing Archimedes, one of the barriers to our understanding of the operational techniques of the Greek geometer is that, although the proof of a difficult theorem may have been discovered first by analysis, only the synthetic proof is formally presented. Often he seems to have preferred the arrangement of the antecedents and consequents in their "natural order," as Pappus has called it.

In actuality, some of the other important methods devised by the Greek geometers, according to later commentators, are intimately connected in spirit at least with analysis. Let us take, for example, the so-called "reduction method." This expression was applied to the general technique of reducing a problem for which the type of proof is not readily apparent to another problem, the solution of which would appear to be possible by known techniques. Thus the difficult problem of the construction of a cube double the volume of a given cube was reduced to what was considered a more easily soluble problem—namely, inserting two mean proportionals between two given lines. Proclus claims that Hippocrates of Chios, possibly the greatest geometer of the fifth century B.C., was the first to use reduction in difficult problems; although this may not be so, it is apparently true that he was the first to use it in the problem of the duplication of the cube. This was one of the three classic problems, the attempted solution of which did much to stimulate the early development of Greek geometry. The other two were the squaring of the circle and the trisection of an angle. Unable to solve these problems by the use of rule and compass, the Greeks were led to construct and study a number of special curves.

The method of reduction applied in these special problems would appear to be analogous on an operational level to formal analysis on a demonstrative level. For if by "reduction" you do not arrive at a proposition or a principle asserted to be true, you do reduce the difficult problem to something that has a more familiar form and thus looks to be soluble.

Still another important methodological concept that took shape in the geometrical corpus was the concept of *diorismoi*, or distinctions determining the possibility of proof. "The *diorismos,*" as a gloss on a work of Pappus has told us, "is a statement in advance as to when, how, and in how many ways the problem will be capable of solution." Proclus further defined the procedure of the *diorismoi* as "the determination of the conditions under which the problem posed is capable of solution, and the conditions under which it is not." The discussion of this concept goes back, in all probability, to the fifth century B.C. The idea of determining the possibility or impossibility of proof has been extremely important in the growth of modern mathematics, particularly in the nineteenth and twentieth centuries; it was less important in the elementary mathematics of the Middle Ages, although on at least one occasion a medieval author attempted to prove the insolubility of a given problem.

In addition to describing the foregoing techniques of proof and solution, we must mention in passing the evolution, during the first three centuries of Greek mathematics, of the general form of presentation that finally appears in Euclid's *Elements*. This form included the over-all organization of a "book" into definitions, axioms, postulates, and propositions and the particular propositional organization embracing the general statement of the proposition, its specification in terms of given geometrical figures, the construction of any additional figures necessary for the proof, and, finally, the proof. We shall have more to say later about Euclid's understanding of these various elements of organization. Now we shall note the importance that formal methods of presentation have as signs, if nothing else, of a high state of mathematical inquiry. Thus early Latin mathematical treatises before A.D. 1000 reveal immediately the poor state of mathematical knowledge by the lamentable perversion and disappearance of the formal Euclidian organization. Conversely, the reappearance on a wide scale of Euclidian and Archimedean forms in the twelfth and thirteenth centuries bespoke a rapidly maturing mathematics and mathematical physics.

As these methodological advances were emerging in the first two or three centuries of Greek thought, some important steps were being taken in the building up of substantial mathematical knowledge. Certain specific theorems were attributed (probably falsely?) to Thales: a circle is bisected by any diameter; in two intersecting straight lines, the opposite angles are equal; etc. Other specific theorems and proofs have been attributed to Pythagoras. The Pythagorean theorem (Euclid, *Elements*, I, prop. 47) is the most famous. We have already mentioned the tradition that claimed Pythagoras as the inventor of abstract and generalized geometry and held that he believed the universe to be numerical in structure. That the Pythagoreans made elementary progress in number theory is to be accepted. They appear to have known and expressed (although by no means invented, as both Egyptian and Babylonian mathematics clearly demonstrate) simple forms of the proportionality of whole numbers, such as arithmetic proportionality where difference in any two successive terms is a constant ($c - b = b - a$) or geometric proportionality where the ratio of any two successive terms, taken in the same order, is a constant. $\left(\frac{c}{b} = \frac{b}{a}\right)$.

The Pythagoreans are also credited with the invention of the method of the application of areas, a method perhaps based ultimately on Babylonian algebra. Its object is the construction of a rectangle or parallelogram equal in area to a given area and having as a base a given line, or a segment of the line, or the line produced. In its simpler forms this method involves the solution of geometric problems equivalent to linear equations in algebra (such as $ax = bc$); solution of the more advanced problems is equivalent to the solution of quadratic equations (expressed in the form $(a + x)\,x = b^2$).

The Pythagoreans may also have discovered that the diagonal of a square is not commensurable with the side—i.e., that there is no common magnitude which taken any number of times will measure both of them exactly. Hence it is believed that the Pythagoreans introduced among the Greeks the study of numerical irrationality. The discovery of irrational magnitudes seems to have been of great importance. Let us suppose that the Pythagoreans really believed that the universe could be described in terms of the natural order of discrete numbers 1, 2, 3, 4, . . . n and the rational fractions formed therefrom. Then where in this order can we place the number representative of the diagonal in terms of the side of a unit square—namely, $\sqrt{2}$? Obviously nowhere, if this series of discrete numbers is considered sufficient, for the $\sqrt{2}$ cannot be expressed in terms of whole numbers and rational fractions. But it can easily be represented in terms of the lines or continuous magnitudes of geometry, where in fact it was discovered. Simply draw a square of side 1, and the length of the diagonal is the desired irrational magnitude. Similarly, the other irrational numbers discovered shortly thereafter could be represented only in terms of the continuous magnitudes of geometry rather than in the discrete numbers of Greek arithmetic. This greater sufficiency of a geometry of continuous magnitudes may well have been one of the decisive factors in its greater growth among the Greeks.

One of the consequences of the discovery of irrationals was that there was then a class of irrational magnitudes not accounted for in the Pythagorean theory of proportionality based on discrete numbers. Tradition has it that it was the mathematicians who surrounded Plato, and particularly his brilliant associate Eudoxus, who fashioned a broader concept of proportionality embracing both rational and irrational magnitudes. Although we have next to nothing extant from Eudoxus, it is believed that the theory of proportionality as presented in the fifth book of Euclid's *Elements* is essentially his.

In spite of the absence of his works the tradition of his greatness is so strong that historians of mathematics are inclined to rank Eudoxus among the very greatest of Greek mathematicians—as one historian says, second only to Archimedes. And if he had a hand in devising the remarkable method of exhaustion that appears in Euclid's *Elements* and that is used with such brilliance by Archimedes, then he deserves this praise. We shall discuss the method of exhaustion presently.

The benefits of the intensive mathematical activity of the fifth and fourth centuries B.C. were reaped by the mathematicians of the Hellenistic period. Among the numerous figures of this period who were important in mathematics, space allows us to mention only three: Euclid, Archimedes, and Apollonius.

Little is known about the life of Euclid, although probably more people have read some version of his *Elements* than have read any other scientific book. We know that he flourished about 300 B.C., or as Proclus says, "in the time of the first Ptolemy." Proclus goes on to repeat one of the stories about Euclid: "and further they say that Ptolemy once asked if there was in geometry any shorter way than that of the elements, and he replied that there was no royal road to geometry." Besides the famous *Elements* of geometry—which Proclus says he put together by "collecting many of Eudoxus' theorems, perfecting many of Theaetetus' [a contemporary of Plato], and also bringing to irrefragable demonstration the things which are somewhat loosely proved by predecessors" —a number of other mathematical and physical works are credited to Euclid. We shall have cause to mention them elsewhere.

As the passage from Proclus suggests, Euclid's great achievement lay not in presentation of original material but rather in his able synthesis of geometrical knowledge. His *Elements* replaced the several earlier collections or "elements" composed by mathematicians of the fourth century. Almost all of us have come into contact in our high-school careers with some form of Euclid's *Elements,* although in most cases this form was much removed from the original. Among the general topics of the thirteen books comprising the *Elements* are plane geometry, the theory of proportions of magnitudes, the nature and properties of whole numbers, and solid geometry. We shall make no attempt to summarize the topics, but the form and organization of the *Elements* are worthy of some note, for such form was quickly adopted in most of the mathematical and physical treatises of the Hellenistic period, and it survived into medieval and modern times. The reader who has looked at the *Mathematical Principles* of Isaac Newton will appreciate its debt to Euclidian form and organization.

In the *Elements* Euclid uses definitions, postulates, axioms, and propositions. Definitions do not constitute a formal part of the proof; they do not state the existence or nonexistence of anything (or if they do, the existence is provisional and awaits formal demonstration) but are descriptions that simply have to be understood. Examples in the first book are the definitions of a line, a plane, and various kinds of triangles. Postulates are unproved but accepted premises. The five postulates of the first book are the most noteworthy, since they really define the whole character of the space and geometry under consideration. The first permits us to draw a straight line from any one point to any other point; the second allows us to extend that line indefinitely from either extremity; the third allows us to draw a circle from any center at any distance from the center; the fourth asserts that all right angles are equal; and the fifth, which is the famous parallel postulate, tells us under what conditions straight lines will intersect. The genius of Euclid in recognizing this fifth postulate as a postulate and not attempting formal proof has often been remarked. It is well known that in the nineteenth century geometries that were not Euclidian were constructed on the basis of denial of one or more of the postulates. Axioms, called by Euclid "common notions," are things that are generally accepted as true and that often apply not only to geometry but to all of mathematics. Thus "things which are equal to the same thing are equal to each other," or "a whole is greater than its parts." Finally, the propositions included by Euclid are of two kinds. The first concern themselves with problems of construction—e.g. (Book I, prop. 1), "to describe an equilateral triangle on a given straight line." The second kind are theorems—e.g. (I, 47), "in any right-angled triangle, the square described on the hypotenuse is equal to the sum of the squares on the other two sides." The organization within propositions was itself very formal and included a number of well-defined steps, such as "Enunciation" and "Example." (Note: For illustration of the formal parts of a "Euclidian" proof, see Appendix II, which gives the Greek names and their English equivalents for a proof of Archimedes.)

A generation or two after the time of Euclid, there lived the most distinguished mathematician of antiquity, Archimedes (*ca.* 287-212 B.C.). It is inferred, but not surely, that he studied at Alexandria with the successors of Euclid. We know him to be a native of Syracuse, where he spent some time in the service of its ruler, Hiero, and where he was slain in its siege by Roman troops in 212 B.C.

His fame as an engine builder is preserved in the fanciful account of Archimedes given by Plutarch in his life of Marcellus, the Roman general who conducted the siege of Syracuse.

> And yet even Archimedes, who was a kinsman and a friend of King Hiero, wrote to him that with any given force it was possible to move any given weight; and emboldened, as we are told, by the strength of his demonstration, he declared that, if there were another world, and he could go to it, he could move this. Hiero was astonished and begged him to put his proposition into execution, and show him some great weight moved by a slight force. Archimedes therefore fixed upon a three-masted merchantman of the royal fleet, which had been dragged ashore by the great labours of many men, and after putting on board many passengers and the customary freight, he seated himself at a distance from her, and without any great effort, but quietly setting in motion with his hand a system of compound pulleys, drew her towards him smoothly and evenly, as though she were gliding through the water. (Plutarch, *Life of Marcellus*, 14, translation of Bernadotte Perin, London, 1917.[2])

Plutarch goes on to describe Archimedes' use of wonderful military engines in the defense of Syracuse, which, however, did not save either Syracuse or Archimedes, for both fell before the Romans. In spite of these great mechanical talents, we are told further by Plutarch, Archimedes held the mechanical arts in disdain, "regarding the work of an engineer and every art that ministers to the needs of life as ignoble and vulgar," and devoted his greatest efforts to theoretical studies, such as geometry. In view of the general noncritical nature

[2] Quoted by permission of the present publisher and holder of the copyright, the Harvard University Press.

of this account there is no good reason to accept this statement as reflecting the actual attitude of Archimedes; nevertheless, it is true that his greatest efforts were in mathematics and in the theoretical and mathematical aspects of statics and hydrostatics. The latter we shall summarize shortly. Here we must say something of his mathematics.

We can glean something of Archimedes' subject matter from a recital of the titles of his works: *On the Sphere and the Cylinder, On the Measurement of a Circle, On Conoids and Spheroids, The Sandreckoner, On Spirals, On the Equilibrium of Planes, On Floating Bodies, On the Quadrature of the Parabola,* and *On the Method.* In much of his mathematical work he concerns himself with finding the areas and volumes of special curved surfaces and solids, relating them to the more readily obtained results for a triangle, a rectangle, a cube, or some such figure. His most powerful tool for proof consisted of (1) a highly developed procedure of reduction to absurdity, combined with (2) the method of exhaustion. The fundamental idea of the method of exhaustion is that one inscribes and/or circumscribes regular figures within or without the figure for which the formulation of the area or volume is sought. Then in regular fashion the areas or volumes of the inscribed or circumscribed figures are increased or decreased until the difference between the unknown area or volume and the inscribed or circumscribed area or volume is *less than any given quantity*. In this method we do *not* say that the increased inscribed figure or the decreased circumscribed figure ever actually reaches the limit—namely, the unknown figure (for this, the Greeks thought, would involve them in fundamental difficulties about the infinitely small). We say only that the inscribed or circumscribed figure can be made to approach *as closely as we like* to the figure under consideration. With this type of exhaustion granted, it can be shown by reduction to absurdity that the area or volume under consideration must correspond to some given formulation, for, if not, a basic contradiction will appear. The reader is invited to inspect in Appendix I the brilliant proof of Archimedes on the area of a circle, in which the method of exhaustion is neatly and simply illustrated.

The reader interested in mathematics may well note one of the difficulties inherent in this use of exhaustion and reduction to absurdity. In a very real sense, one must know the answer before he begins. He assumes a given formulation and then shows that if it is not true a contradiction follows. But where does he get the given formulation? Asking this question highlights one of the fundamental problems in the history of science, *the distinction of method of discovery from method of proof*. To understand how science has developed and how scientists have worked we want to know how the scientist first gets the answers as well as how he proves them. Unfortunately the documents rarely tell us of the almost intuitive methods by which basic problems are solved. But in the case of Archimedes we are lucky to have extant a treatise rediscovered scarcely more than a generation ago, entitled *On the Method*. It illustrates a mechanical method by which Archimedes mathematically balanced geometrical figures as on a weighing balance. He balanced a figure the area or volume of which was desired against one the area or volume of which was known; then, by noting the relative distances of the centers of gravities of the two figures from the fulcrum of the balance and by applying the law of the lever, he arrived at the desired area or volume in terms of the known area or volume (see Appendix II). One of the basic assumptions employed in this method is that an area can be considered a summation of an infinite number of line segments or a volume an infinite number of plane segments. Archimedes was quite aware, as his introductory letter tells us, that his method was not rigorous so far as demonstration was concerned, but it allowed him to find the answer and thus apply the more rigorous method of exhaustion and reduction to absurdity:

> . . . certain things first became clear to me by a mechanical method, although they had to be demonstrated by geometry afterwards because their investigation by the said method did not furnish an actual demonstration. But it is of course easier, when we have previously acquired, by the method, some knowledge of the questions, to supply the proof than it is to find it without any previous knowledge . . .

and I deem it necessary to expound the method partly because I have already spoken of it . . . but equally because I am persuaded that it will be of no little service to mathematics; for I apprehended that some, either of my contemporaries or of my successors, will, by means of the method when once established, be able to discover other theorems in addition, which have not yet occurred to me. (Archimedes, *Method,* Introduction, 2, translation of T. L. Heath, Cambridge, Eng., 1912.[3])

By the elegance of his methods Archimedes was able to demonstrate theorems equivalent to the formulations $S = 4\pi r^2$ for the surface of a sphere and $V = \frac{4}{3}\pi r^3$ for its volume, as well as that the segment of a parabola is equal to four thirds the triangle having the same base and altitude as the segment. It has often been pointed out that in some of his more elegant theorems "he performs the equivalent to the operations of integration" in integral calculus. Of considerable interest in the history of mathematics is his system of representing large numbers, outlined in *The Sandreckoner,* and his determination of π as $3\frac{10}{71} < \pi < 3\frac{1}{7}$.

But what strikes the historian of science as the most important achievement of Archimedes is his extraordinary influence in the development of a mathematical and quantitative view of physical problems in the sixteenth and seventeenth centuries, an influence that had already been previewed by the spread and use of some of his works in the thirteenth century. The influence of Hellenistic geometry in general and of Euclid and Archimedes in particular in the high and late Middle Ages is a topic that will receive the author's attention in a later volume.

We can conclude our brief summary of Hellenistic mathematics with passing reference to the third of the great triumvirate, Apollonius of Perga, who flourished at the end of the third and the beginning of the second centuries B.C. Of the eleven works attributed to him, the most important (and one of the two to survive) is his *Treatise on Conic Sec-*

[3] Quoted by permission of the publisher, the University Press, Cambridge, England.

tions, which deals, of course, with theorems relative to ellipses, parabolas, and hyperbolas. It represents the culminating effort of some two centuries or more of study of conics that grew out of attempts to solve such classic problems as that of the duplication of a cube (see Appendix III for some remarks on conic sections). Its influence both upon Arabic mathematicians and, later, upon early modern scientists such as Galileo and Newton was great.

We are not to think that Greek mathematics immediately declines with the passing of Euclid, Archimedes, and Apollonius or that its study is limited to them. Only the three greatest of the Greek mathematicians have been selected for the present discussion. Greek mathematics continued to be studied and written about during the whole late antique period, and we shall encounter later the names of Hero of Alexandria, Diophantus, Nicomachus, Pappus, and Proclus, as we study its continuing tradition.

Chapter 6

Greek Physics

WE DO not have to wait until medieval or early modern times for the application of geometry to the investigation of nature, for it began in both physics and astronomy in the fourth century and matured in the Hellenistic and Greco-Roman periods. We have already suggested the basic importance of the Pythagorean mathematical point of view of nature; but when the mathematical view was coupled with something close to scorn of the world of the senses, as it was in some of the Platonic dialogues, little sound physics could arise. Even the most apologetic Platonist will not stand behind Plato's *Timaeus* as a work of high scientific caliber, although it is true that some of the ideas suggested therein were not without their influence on Aristotle and later authors. On the other hand, the *Physics* of Aristotle, in spite of its general nonmathematical and in a sense qualitative character, was enormously influential in the subsequent development of physical ideas during the medieval period. Here we shall briefly summarize some of the most important views in the *Physics,* to which we shall return in Chapter 13 when discussing Aristotelian commentaries in late antiquity.

Aristotle's physical picture distinguishes between celestial and terrestrial areas of activity. Looking for the moment at the terrestrial world, we are aware first of the doctrine of matter and form already mentioned (Chap. 3). There is a substratum, a prime matter, the independent existence of which is only potential but the actual existence of which is always in conjunction with form. Something is said to belong to classes as a result of its form, but it is said to be individual as the result of matter. In short, matter is the principle of individuation. Form gives something its essential character. It is by its form that we recognize it for what it is. But when we examine more deeply the principles of things we are struck by four types of causation —one might say by four "factors" that are involved in things that exist in the terrestrial world. The first is the "material" factor, the second is the "formal," the third is the efficient, and the fourth is the final or purposeful cause. The four causes can be explained by analogy with something artificially produced. A bed is a bed because it is made of wood (the material cause), in a given shape (its formal cause), by a carpenter (its efficient cause), for the purpose of providing slumber (final cause).

As we look at the fundamental manifestations of matter and form in the terrestrial world we detect or infer (for they are never in a pure state) the existence of four basic elements—earth, water, air, and fire—each having a distinctive pair of qualities. Earth is cold and dry; water, cold and wet; air, hot and wet; fire, hot and dry. These elements tend to arrange themselves concentrically about the center of the world: the earth is a sphere at the center, and the water, air, and fire are successive shells about the core of the earth. There is constant change as the elements are transformed into one another when the substratum is differently formed, or as the elements compound themselves differently with one another. In the elementary changes, the substratum is common to the four elements. The commentators later suggest that there is some prime form joined with prime matter (and as potential as the prime matter) which constitutes corporeity, also common to all of these elements, and which renders matter susceptible to receiving three dimensions. Now, as we have suggested, the elements tend to seek a static arrangement according to their places about the center of the world, but because of the continuous change in elements and compounds—produced ultimately by the everlasting movement of the celestial spheres—in actuality change and movement are the order of nature. The study of nature becomes the study of movement.

Movement in its broadest sense is said by Aristotle in some passages to include not only locomotion—i.e., change of place—but also qualitative alteration, quantitative augmentation and diminution, and even on occasion generation (coming into being) and corruption (passing away). In an effort to find a definition to include all these categories, he seizes upon a description of movement as the actualizing of something that exists in potentiality. For example, something white undergoes qualitative alteration and becomes black. When actually white, it nevertheless is potentially its contrary, black. This particular qualitative movement is, then, the actualizing of the potential blackness. Such a definition seems to be an ambiguous locution, for the nature of *potential existence* is most elusive. Essential to this concept of movement was the doctrine of the existence of fundamental sets or pairs of contraries, with movement being the passage from one contrary of the pair to the other. This approach to movement as the actualizing of the potential seems to have been advanced to get around the basic criticism of the Eleatic philosophers of the preceding century that change and movement do not exist, for if they did, something existent would follow from something nonexistent, or in short the nonexistent must exist, a fundamental contradiction. Aristotle is saying rather that movement and change are not examples of the existent arising from the nonexistent but only of changes of the mode of existence, changes from potential existence to actual existence.

Local movement of the elements and their compounds—i.e., terrestrial local movement—is governed by the doctrine of natural place. In the concentric arrangement of the elements each element has its natural place, the earth at the center, the water adjacent to it, and the air and fire in successive places. Now if an element is removed from its natural place, it tends to return to that place by a straight-line movement. Hence if we pick up a stone and let go of it, since it is predominantly earth it falls downward in an attempt to return to its natural place near the center of the universe. The same is true with water; rain generated in the midst of the air tends to seek its natural place and so falls. Air and fire, on the other hand, tend upward, for their natural place is above us. The movement resulting from the tendency of elements to return to their natural places is called *natural movement*. Contrariwise, if we remove a body from its natural place—e.g., lift a stone—we have acted against the natural tendency of the body, producing thereby unnatural or *violent movement*.

It can be seen that local movement is intimately tied to qualitative changes or alteration, for if a transformation of elements, one into another, takes place and the new element is formed out of its natural place, then the tendency of that newly formed element will be to seek its natural place, thus producing movement. Similarly, if the proportion of elements in a mixture or a compound is altered sufficiently, movement can result. Thus something that is predominantly earth may by exposure to heat or some other influence change to a mixture predominantly of fire, and the result is the motion of the new mixture upward. And since both elementary and mixture changes are going on unceasingly, the terrestrial world is essentially in motion. The laws of movement that Aristotle deduces from gross experience, laws that necessitate the continual presence of mover and moved, force and resistance, for the continuance of movement, we shall deal with in Chapter 13, when we take up their criticism by late antique authors. We can observe here that they are of only a quasi-quantitative nature.

We have spoken briefly of the terrestrial, or sublunar, world according to Aristotle. Distinct from it, but surrounding it and influencing its complex of movement and change, is the finite celestial world, including the moon, the sun, the planets, and the stars. Here the laws of movement and the nature of the constituent element differ from those in the sublunar world. Celestial bodies are formed of a fifth element, ether (the later quintessence); and ether is unchanging and eternal so far as its qualitative nature is concerned. This celestial fifth element possesses only the tendency for uniform, continuous, circular movement. Thus circular movement is natural to celestial bodies and actual movement is presumably engendered by souls, although the nature of the movers in the case of the various celestial motions is much mooted. Unlike sublunar rectilinear movement, the circular movement of the celestial bodies does not require the continued presence and substantial contact of both force and resistance.

The distinction between the sublunar and celestial regions, with the consequent distinction of the physical laws to describe each, although completely rejected by atomists throughout our period, kept a tenacious hold on a considerable majority of natural philosophers until the sixteenth century.

Aristotle's rejection of an actual infinite, in the form of either the infinitely large or the infinitely small, his denial of the existence of a void, and his consequent definition of place and space in terms relative to contiguous bodies are characteristic Aristotelian doctrine to which we shall return later.

Before leaving Aristotle we ought to note that his is the first *extant* discussion of the kinematic aspect of movement. He is the first one whose treatise we have to give us rules for comparing the speeds of bodies in terms of the space traversed in given times. Now Aristotle does not tell us neatly and generally that $V = \frac{S}{T}$ where V is a velocity, S is a distance and T is a time. Rather, he gives us a whole set of specific comparisons. In one chapter he tells us what is meant by the term "quicker." Here he says that the movement of one body is said to be quicker than that of another in three ways. These can be represented symbolically as follows: Assume that V_1 and V_2 represent the "movements" or different speeds of two bodies. Now assume that the body with a speed of V_1 traverses distance S_1 in time T_1, and that the body with a speed of V_2 traverses distance S_2 in time T_2. Then, according to Aristotle, three statements can be made about the relation of the one movement to the other. These are: (1) $V_2 > V_1$ if $S_2 > S_1$ when $T_2 = T_1$; (2) $V_2 > V_1$ if $S_2 = S_1$ and $T_2 < T_1$; (3) $V_2 > V_1$ if $S_2 > S_1$ and $T_2 < T_1$. In another passage he notes that the movement of one body can be said to be equal to that of another body when the same distance is traversed in the same time; or, symbolically: $V_2 = V_1$ if $S_2 = S_1$ and $T_2 = T_1$. Now we can derive from all four of these statements our general expression $V = \frac{S}{T}$ once we have decided that velocity can be represented as a ratio, but Aristotle does not go ahead to the general statement of the simple definition of speed as a ratio of distance over time. (In fact Greek geometers generally do not express their quantification of variables and relationships in nature by means of metrical formulae involving the ratios of unlike quantities but rather by means of mathematical statements involving true proportions of like quantities, such as distance is to distance as time is to time, etc.) No doubt the whole analysis of speed in terms of distance and time had been initiated even earlier, in the fifth century B.C., by the developing interest in geometry and in special kinds of curves that were constructed kinematically. To give an example of this kind of approach to geometry we could say that a circle could be constructed kinematically if we allow a line to rotate in a plane about one of its termini. Many higher curves were constructed only in kinematic terms—that is to say, by the movements of points or lines. No doubt the interest in astronomy, which centered in the problem of a geometrical representation of the movements of heavenly bodies, also stimulated the interest in kinematics.

First results of the investigations into physical phenomena undertaken by Aristotle at the Lyceum appeared in the generation or two after Aristotle's death. Among them were the following: (1) the activity of Strato the Physicist (*fl.* 287 B.C.), whose views often differed radically from those of Aristotle; (2) the writing of a treatise called *Mechanics,* which was afterwards ascribed to Aristotle but which may have been written by Strato or some contemporary; (3) and the preparation of the treatise *On Audibles,* ascribed to Aristotle, but perhaps by Strato.

Strato holds an important place in this brief description of Greek science, for he apparently represents a link between the Lyceum at Athens and the great Museum at Alexandria. Little is known of his life. He was born at Lampsacus and no doubt studied at the Lyceum. He is mentioned as a tutor of the son of the first King Ptolemy of Egypt. We know that in 287 B.C. he was called to succeed Theophrastus as head of the Lyceum in Athens. Over forty writings have been attributed to him by later authors, but all are lost except for fragments. They include writings on logic, ethics, metaphysics, physiology, psychology, and, above all, physics. Polybius, Cicero, and Simplicius call him Strato the Physicist (the Natural Philosopher), no doubt to distinguish him from a namesake, Strato the Physician.

Strato had an apparently deserved reputation as an experimenter. The passage summarizing Strato's idea on the corporality of air and the existence of a vacuum, which is quoted from Hero (Chap. 2), is a clear case of the use of experiment as an active control of natural phenomena to confirm scientific hypotheses. Strato is usually represented as assuming a compromise position between the atomists and Aristotle. Like Aristotle he seems to have held that no natural continuous vacuum existed. But he asserted that a continuous vacuum could be produced artificially, as experiment clearly proves. At the same time, experiment showed the existence of fine vacua distributed between the particles of bodies, a fact reflective of the atomic position. But, probably aware of the necessity of close liaison with experience, Strato is said to have been amused by the hooked character attributed to atoms by Democritus, since experience neither confirms nor denies this kind of assumption.

Interesting also are Strato's views on weight. We know from Cicero that Strato taught that everything which is or which is made is the result of weights and movements, a position which, according to Cicero, "frees God from his great work and me from fear." Unlike Aristotle (but like Epicurus, according to Simplicius) Strato believed that all bodies have weight and tend toward the center, but the lighter are displaced and forced upward by the pressure of the heavier ones. With this correct view, he set aside the artificial distinction of heavy and light bodies made by Aristotle.

Strato's views on movement, so far as we know them, are important. We know from a passage from the later commentator Simplicius (sixth century) that Strato composed a work, now lost, called *On Motion*. Simplicius quotes from this work in connection with a discussion regarding acceleration:

> It is universally asserted as self-evident that bodies moving naturally to their natural places undergo acceleration. . . . But few adduce any proof of the fact itself. . . . It may not therefore be out of place to set forth the indications (of acceleration) given by Strato, the Physicist. For in his treatise *On Motion*, after asserting that a body so moving completes the last stage of its trajectory in the shortest time,

he adds: "In the case of bodies moving through the air under the influence of their weight this is clearly what happens. For if one observes water pouring down from a roof and falling from a considerable height, the flow at the top is seen to be continuous, but the water at the bottom falls to the ground in discontinuous parts. This would never happen unless the water traversed each successive space more swiftly. . . ."

Strato also adduces another argument as follows: "If one drops a stone or any other weight from a height of about an inch, the impact made on the ground will not be perceptible, but if one drops the object from a height of a hundred feet or more, the impact on the ground will be a more powerful one. Now there is no other cause for this powerful impact. For the weight of the object is not greater, the object itself has not become greater, it does not strike a greater space of ground, nor is it impelled by a greater (external force). It is merely a case of acceleration. And it is because of this acceleration that this phenomenon and many others take place." (M. Cohen and I. E. Drabkin, *A Source Book in Greek Science,* New York, 1948, pp. 211-212.[1])

One of the most interesting features of this passage and one that is generally overlooked is that it seems to approach accelerated movement from a kinematic rather than from a dynamic point of view; i.e., Strato seems to be thinking of acceleration in terms of the dimensions of space and time rather than in terms of the forces producing acceleration. We have already seen that Aristotle gave us some rudimentary rules for determining which of two movements is quicker. Furthermore, Aristotle distinguished uniform velocity (*isotachōs*) from non-uniform. Non-uniform velocities were classified as increasing or decreasing speed (". . . it makes no difference whether the velocity is increased, decreased, or is uniform"). A movement is non-uniform because its parts are unequal, i.e., varying times for equal spaces or varying magnitudes of distance in equal periods of time. With Strato the analysis appears to be

[1] Quoted by permission of the publisher, the McGraw-Hill Book Company, Inc.; copyright, 1948.

more explicit. Like Aristotle he applies the concept of the "quicker" to the parts of a single movement. Thus he gives us the first analysis (even if it is incomplete) of acceleration in terms of space and time. We can represent Strato's analysis in modern terms as follows: The statement describing acceleration by the completion of the last part of a trajectory in the shortest time assumes that (1) for distances $S_1 = S_2 = S_3 = \dots S_n$, the times are related as follows: $t_1 > t_2 > t_3 > \dots t_n$. He then goes on to conclude that acceleration assumes (2) that for $S_1 = S_2 = S_3 = \dots S_n$, the speeds are related as follows: $V_1 < V_2 < V_3 \dots < V_n$. One can get from (1) to (2) by applying Aristotle's second definition of the "quicker" —namely, that which traverses the same space in less time. In summary, the definition of acceleration understood by Strato would be movement such that equal spaces are traversed in succeeding periods of less time—i.e., at continually greater speed. If we assume continuous variation (i.e., if we assume, as Aristotle and Strato did, that both distance and movement were continuous quantities marked by divisibility to infinity), then Strato seems to be saying that, for falling bodies, speed is directly proportional to the distance of fall ($V \propto S$). The correct description is, rather, that speed is directly proportional to the time of fall, but we must await late medieval and early modern times for the correct enunciation of the law of free fall. Presumably the third-century commentator on Aristotle, Alexander of Aphrodisias, was following Strato when he claimed that "bodies move downward more swiftly in proportion to their distance from above."

All in all, it is to be lamented that Strato's *On Motion* is lost, for perhaps he carried this analysis even further than the brief quotation of Simplicius suggests. It can be observed further that, so far as we know, no other figures in antiquity treated the problem of acceleration from this kinematic point of view, although many discussed the causes of acceleration. It is possible, of course, that the whole problem was treated both dynamically and kinematically by the famous astronomer of the second century B.C., Hipparchus, in a now lost treatise entitled *On Bodies Carried Down by Weight*. But so far as the evidence that we have now is concerned we must conclude that it was not until the thirteenth and fourteenth centuries in the Latin West that we get a detailed analysis of uniformly accelerated movement.

To return to Strato, we should note that there is a tradition that the famous astronomer Aristarchus, the so-called "Copernicus of antiquity" (Chap. 7), was a student of Strato. If this is true, it connects Strato solidly with the activity of the Museum at Alexandria. It has also been suggested that the views of the physician Erasistratus reflect Strato's ideas on the vacuum and that the great physiologist's experimental approach may stem from Strato's influence.

The second evidence of physical investigation in the period of transition between the high point of the Lyceum at Athens and the Museum at Alexandria is the Aristotelian treatise *Mechanics* (or, more accurately, *Mechanical Problems*), which, as we have suggested, may have been written by Strato or one of his contemporaries. Apparently formal mechanical treatises go back at least to the beginning of the fourth century B.C. One of Plato's contemporaries, Archytas of Tarentum, is said to have written "the first systematic treatise on mechanics based on mathematical principles." If this is so, then it would be difficult to avoid concluding that the law of the lever and other fundamental mechanical principles such as appear in the later mechanical treatises were contained in Archytas' work in some germinal fashion. The Aristotelian *Mechanics,* which seems to be our first extant treatise of this type, already contains the law of the lever, the statement of the inverse proportionality of the weights and the lever arm lengths. But there is no formal mathematical demonstration of the law, as in the *Treatise on the Balance* attributed to Euclid or as in Archimedes' *On the Equilibrium of Planes,* where one of the most elegant proofs in antique physical treatises is found. The author of *Mechanics* explains the law of the lever by stating that if movement has been imparted to a lever in equilibrium, the velocities of the weights—i.e., the arcs swept out in equal time—will be inversely proportional to the suspended weights. At least one important historian of ancient physics, P. Duhem, has connected this exposition of the law of the lever with the general Aristotelian laws of movement and has held that it contains in germinal form the "dynamic" approach to statics that in modern times was to replace the "static" approach of Archimedes. This may very well be true, but we must not lose sight of the fact that there is an extraordinary gulf in scientific maturity between the rather confused statements of the *Mechanics* and the mathematically sophisticated proofs of Archimedes.

Another interesting theory exposed in the *Mechanics* is that of the so-called "parallelogram of velocities." This enunciation seems to represent the first mathematical exposition of this important principle of composite movements. It is proved by easy geometrical method that when a body is possessed of two uniform velocities differently directed, the resultant velocity can be represented by the diagonal of the parallelogram of which the adjacent sides represent the magnitudes (and directions) of the two velocities. The author of the *Mechanics* also realized, without providing an extended analysis, that if the ratio of the two velocities was not a constant (or if the two velocities were not both uniform), then the line representing the composition or resultant of the two velocities would not be a straight line.

Also of interest in the treatise is the nice analysis of the horizontal stability of a lever supported from above. Wedges, pulleys, and the steelyard (the balance of unequal arm lengths) are treated by the author. And he asks the searching question, "Why is it that a body which is already in motion is easier to move than one at rest?" The answer, he believes, lies in the fact that a body at rest has a contrary resistance to being moved not so great as, but similar to, the resistance presented by a body moving in a direction contrary to that of the motivating force.

We have posed as the third important post-Aristotelian development in physics the writing of the treatise on sound entitled *On Audibles*, which, although attributed to Aristotle, is probably by Strato or some other figure of the Lyceum (some scholars have even assigned it to Heraclides of Pontus). Interest in sonic phenomena did not, of course, begin with this treatise. We have already mentioned that Pythagoras or the early Pythagoreans investigated the relationships of pitch with string lengths (see Chap. 2). Thus an author of the second century, Theon of Smyrna, tells us that Pythagoras "investigated these ratios [i.e., the fourth, the fifth, the octave, etc.] on the basis of the length and thickness of strings, and also on the basis of the tension obtained by turning the pegs or by the more familiar method of suspending weights from the strings" (Cohen and Drabkin, *A Source Book in Greek Science,* p. 295). As to the physical nature of sound, we are told in a fragment of a work of an early Pythagorean, Archytas, that there cannot be sound "without the striking of bodies against one another," and further that swift motion produces a high-pitched sound and slow motion one at low pitch. Aristotle, in his treatise *On the Soul,* also indicated that it is the colliding of bodies with one another and the air that produces sound, and he recognized that a medium such as water or air is an indispensable condition of hearing: ". . . and it is the air which causes hearing, when being one and continuous it is set in motion. . . . That, then, is resonant which is capable of exciting motion in a mass of air continuously one as far as the ear. There is air naturally attached to the ear. And because the ear is in air, when the external air is set in motion, the air within the ear moves." (Translation of R. D. Hicks, Book Two, 8). In the treatise *On Audibles* we are told even more explicitly how the air transmits sound: "For, when the nearest portion of it is struck by the breath which comes into contact with it, the air is at once driven forcibly on, thrusting in like manner the adjoining air, so that the sound travels unaltered in quality as far as the disturbance of the air manages to reach" (Oxford translation of Loveday and Forster, 1913, of 800 a1-b3). Much later, in the fifth century, Boethius, following the Greek tradition, was to spell out even more explicitly a nascent "wave theory" of sound as he connected frequency of vibration with pitch:

> 14. Let us now speak of the method of hearing. In the case of sounds something of the same sort takes place as when a stone is thrown out and falls onto a pool or other calm water. The stone first produces a wave with a very small circumference. Then it causes the waves to spread out

in ever wider circles until the motion, growing weaker as the waves spread out, finally ceases. The later and larger the wave, the weaker the impulse with which it breaks. Now if there is an object that can block the waves as they grow larger, the motion is at once reversed and forced back, in the same series of waves, to the center from which it originated.

In the same way, then, when air is struck and produces a sound, it impels other air next to it and in a certain way sets a rounded wave of air in motion, and is thus dispersed and strikes simultaneously the hearing of all who are standing around. And the sound is less clear to one who stands further away since the wave of impelled air which comes to him is weaker. (Cohen and Drabkin, *A Source Book in Greek Science*, pp. 293-294.[2])

But let us return to the post-Aristotelian physics. The highest development of the mechanical tradition inaugurated in the fourth century B.C. took place in the Hellenistic and post-Hellenistic periods. To Euclid is attributed by Islamic and later Latin scribes certain mechanical writings, or rather, fragments of writings. Two of these are directly Aristotelian in character. The third, entitled *Treatise on the Balance,* may be genuine and is of interest because, unlike the statement of the law of the lever in the Aristotelian *Mechanics,* its statement on the subject is proved on entirely geometrical grounds. It clearly presents the basic idea of static moment, that effective force of a weight in a lever system is measured by the product of the weight and the horizontal distance of the vertical line running through the fulcrum, regardless of what angle the weights' suspension line makes with the lever arm.

Like Euclid's proof of the law of the lever, that of Archimedes in the *On the Equilibrium of Planes* is based completely on statical-geometrical grounds (see Appendix IV). The proof of Archimedes depends fundamentally on the extension of two ideas: (1) equal weights at equal distances from the fulcrum are in equilibrium (postulate 1); and (2) the center of

[2] Quoted by permission of the publisher, the McGraw-Hill Book Company, Inc.; copyright, 1948.

gravity of two equal weights not having the same center of gravity is at the middle point of the line connecting their individual centers of gravity (proposition 4). At the heart of the brilliant proof of the law (propositions 6 and 7), then, is an appeal to the geometrical symmetry apparent in the fundamental ideas noted. The proof is among the ablest that Archimedes has given us.

Archimedes utilized mathematics in a completely mature fashion for the treatment of a physical problem. We should note the abstraction evident in the Archimedean demonstration. He has dispensed with all the properties of weights not bearing on the mechanical theorem being proved. His weights have become *idealized geometrical magnitudes* not actually and completely realizable. This mathematical abstraction is to be emphasized, for it was to influence Galileo and his successors—and in fact, his predecessors in the thirteenth and fourteenth centuries, who were as early proponents of kinematics to speak of points, lines, and magnitudes in motion.

To problems of hydrostatics Archimedes in his *On Floating Bodies* applies the same kind of geometrical analysis. Thus, on the basis of two geometrically expressed initial propositions, Archimedes demonstrates propositions concerning the relative immersion and weight in a fluid of solids less dense than, as dense as, and more dense than the fluid. The proposition relative to solids more dense than the fluid (proposition 7) expresses the so-called "principle of Archimedes"—namely, that "the solid will, when weighed in the fluid, be lighter than its weight in air by the weight of the fluid displaced." Book II, which takes up the stability of floating paraboloid segments, is even more geometrical in character. Archimedes' contributions to hydrostatics are reflected on the legendary side by the widely known bathtub incident related by the Latin writer Vitruvius. We are told that, as Archimedes entered the tub, the more he submerged his body, the more the water spilled over the tub, and this led him suddenly to the method of determining whether the crown being made for King Hiero was pure gold, for if it were not the crown would displace a greater volume of water than a weight of gold equal to the weight of the crown, since the equal weight of gold would have a greater density or specific gravity and thus occupy less volume than the crown made of alloy. At any rate, according to Vitruvius, Archimedes ". . . without a moment's delay and transported with joy . . . jumped out of the tub and rushed home naked, crying out in a loud voice that he found what he was seeking; for as he ran he shouted repeatedly in Greek, 'Eureka, eureka.' "

The tradition of formal mechanical treatises, including both theoretical and applied mechanics as distinct from the almost completely theoretical mechanics of Archimedes, was continued in the third century B.C. by Ctesibius (*fl. ca.* 283-247 B.C.?), whose work is lost and is known only by the references of Vitruvius (see Chap. 8, § IV), and by Philo of Byzantium (*fl. ca.* 250 B.C.?), who wrote a composite *Mechanical Syntaxis,* of which only the fourth part, on artillery machines, and the fifth part, on pneumatic phenomena and machines, are extant. Unfortunately the part dealing with simple machines such as the lever, which undoubtedly contained something of theoretical mechanics, is missing. But in the artillery section there are worked ballistic formulae of an empirical character, and the section on pneumatics contains the description of a thermoscope, which, although it has the avowed purpose of illustrating the relationship of fire and air, at the same time represents differences in temperature.

The most comprehensive of the extant mechanical treatises is that of Hero of Alexandria (first century ?), who was at the same time a fine mathematician and an excellent mechanician. In his *Mechanics* he kept in mind the theoretical basis of the machines described. In the second section of his work he described five simple machines "by the use of which a given weight is moved by a given force." These machines were the wheel and axle, the lever, a system of pulleys, the wedge, and the screw. As Pappus of Alexandria (end of third and beginning of fourth centuries) remarks, Philo and Hero both reduced these machines to a single principle. Today we say that the ratio of the motivating force to the weight moved is inversely proportional to the ratio of the distance through which the motivating force acts to the distance through which the weight is moved. This idea is expressed in another way by

Hero: the smaller the force we have available to lift a given weight, the longer it will take us. Suppose we have the axle-wheel arrangement shown in Figure 8. A weight of 1000 talents is hung on axle A. Wheel B has a radius five times that of axle A. Hence a weight or force of 200 talents applied to the rim of B will just equalize the 1000 talents at A, or a

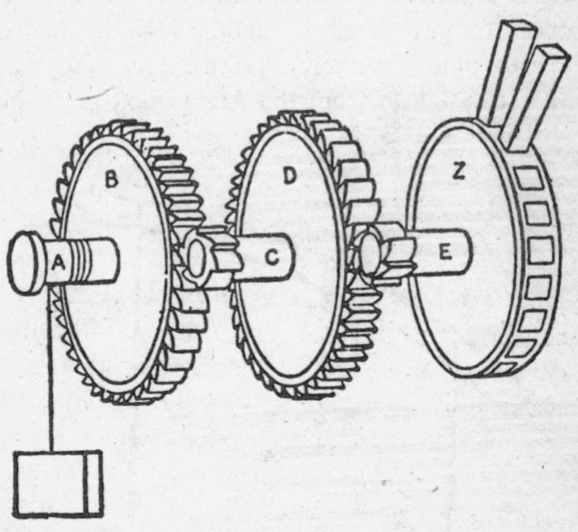

FIG. 8. An axle-wheel train, according to Hero.

weight of 200 plus will lift the 1000 talents. B is geared to axle C, which is equal in radius to axle A. C is fixed to wheel D, which is equal in radius to wheel B. Hence a weight of 40 talents on the rim of D will just balance the 1000 talents at A, and a weight of 40 plus will just move the 1000 talents. Now suppose by the application of 40 plus at the rim of D we turn D through a single revolution, and suppose that by the application of 200 plus on B we turn the equal wheel B one revolution. Now a single revolution of D lifts the 1000 talents only one fifth of the distance that a single revolution of B does. Hence, if we are going to lift the 1000 talents by the 40 plus talents at D rather than by the 200 plus talents at B, we must turn D through five revolutions for every revolution of B, and thus it will take us five times as long using the 40 plus talents at D as using the 200 plus talents at B. Hero expresses the general principle as follows:

> In this machine (wheel and axle) and all the machines like it which are productive of great force (*quwat kabīrat*) there is retardation, because in the proportion that the moving force which moves the large weight is weak, so in this proportion it must be extended in time. For just as is the proportion of one force to another, so is that of time (*zamān*) to time [inversely]. (Hero, *Mechanics,* Book II, Chap. 22, translation from the Arabic text of L. Nix.)

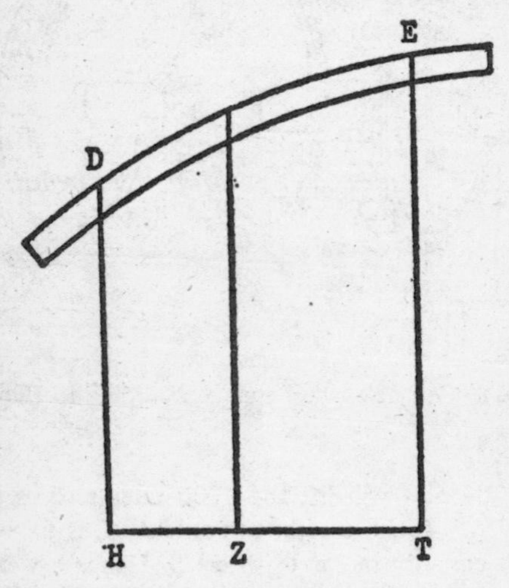

FIG. 9. The bent lever.

The same principle expressed in the same words with a similar explanation is also presented by Hero for a system of pulleys and for the lever. This is the fundamental principle called later the principle of virtual velocities or virtual work, and its recognition is a great achievement of the Greek mechanicians.

Finally, we should note that Hero in the *Mechanics* (Book I, Chap. 33) generalizes the principle of the lever further, applying it to the bent lever; thus he says that the ratio of the weights is equal to the inverse ratio of the horizontal distances to the vertical through the fulcrum, rather than to the inverse ratio of the lengths of the bent lever arms. Referring to the accompanying figure (Fig. 9) he declares that the ratio of the weights hung from D and E is inversely as the ratio of HZ to ZT.

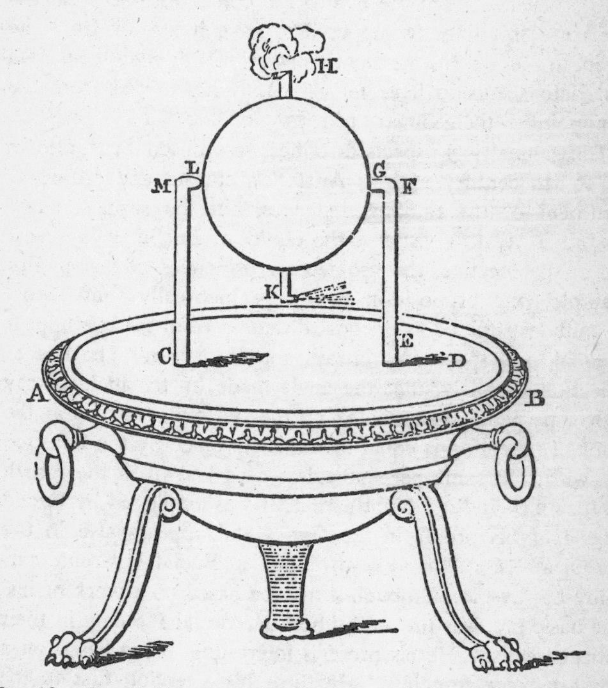

FIG. 10. Hero's "steam engine": a ball rotated by the "jet" action of steam.

The numerous mechanical contrivances as given in another work by Hero, entitled *Pneumatics*, have always drawn curious attention, particularly the so-called "Hero's steam engine," which used the propelling force of two steam jets to produce the rotary motion of a glass ball mounted on an axis (see Fig. 10).

Our account of Greek physics would not be complete without some reference to the optical investigations made during the Hellenistic and Greco-Roman periods. We have already drawn attention, in Chapter 2, to the fact that optics among the Greeks was a discipline of a distinctly experimental and mathematical character. The science of optics was in a sense built up independently of the theories of vision. Whether visual rays proceeded from the eye to the object (as the Pythagoreans are traditionally supposed to have held), or from the object to the eye (as the atomists asserted), or even from both (as Plato seems to have indicated), is immaterial from the standpoint of the geometry of rays.

The geometrical aspects of optics were no doubt studied in the fourth century B.C., as Aristotle's curious and erroneous treatment of the rainbow indicates, but the earliest extant treatise so treating vision is the *Optics* of Euclid. It is a treatise on perspective, the geometrical principles of vision. For example, one proposition proves geometrically that "equal magnitudes situated at unequal distances from the eye appear unequal, and the nearer always appears larger." The law of reflection—namely, that the angle made by the incident ray with a perpendicular erected on the reflecting surface at the point of reflection is equal to the angle made by the reflected ray with that same perpendicular—was known in the fourth century B.C. and no doubt earlier. It was employed by Euclid in one of his proofs in the *Optics* and appears also in the *Catoptrics* (*On Mirrors*), attributed to Euclid but quite certainly not by him, although it may be based on a work of his. The basic law was proved by both Hero and Ptolemy in their optical treatises. Hero's proof is interesting, being based on a "least-distance principle." He turns his attention first to the straight-line propagation of light, which rests on the fact that the rays "are emitted with infinite (in the sense of enormously large) velocity. Therefore they will suffer neither interruption nor curvature, nor breaking, but will move along the shortest path, a straight line." Then, proceeding to the basic reflection law, he assumes that the path of the rays from eye to mirror to object is a minimum, and he easily demonstrates by geometry that the minimum path results when the angle of incidence equals the angle of reflection.

Ptolemy, on the other hand, in his *Optics* (if indeed he is the author of this work) gives experimental confirmation of the reflection law as well as of a number of other basic optical principles. But even more interesting are his experimental efforts in connection with the refraction of light passing from air to water, air to glass, and water to glass. He notes values of the angle of refraction—i.e., the angle made by the ray as it emerges into the other medium with the perpendicular at the point of refraction for a whole series of incident angles increasing by 10°. Although he realized "that the angles (of incident and refraction) as measured from the perpendicular have a definite quantitative relationship," Ptolemy failed, as indeed did all the antique and medieval optical investigators, to find that relationship, the so-called sine law described by both Snell and Descartes in the beginning of the seventeenth century. By study of his tables it has been inferred that Ptolemy believed the following relationship to hold between the angle of incidence (i) and the angle of refraction (r): $r = ai - bi^2$, where a and b are constants that vary for the different media. This is not as close as the sine law, $\frac{\text{sine } i}{\text{sine } r} = k$; but it is not very far off, and it is better than $\frac{i}{r} = k$, which also was occasionally affirmed.

Ptolemy's entire experimental procedure can be illustrated by the following extract. It is quoted at length, for it set the pattern for the standard medieval optical experiments.

> The amount of refraction which takes place in water and which may be observed is determined by an experiment like that which we performed with the aid of a copper disk, in examining the laws of mirrors.
>
> On this disk draw a circle (see Fig. 11) ABGD with center at E and two diameters AEG and BED intersecting at right angles. Divide each quadrant into ninety equal parts and place over the center a very small colored marker. Then set the disk upright in a small basin and pour into the

basin clear water in moderate amount so that the view is not obstructed. Let the surface of the disk, standing perpendicular to the surface of the water, be bisected by the latter, half the circle, and only half, that is, BGD, being entirely below the water. Let diameter AEG be perpendicular to the surface of the water.

Now take a measured arc, say AZ, from point A, in one of the two quadrants of the disk which are above the water level. Place over Z a small colored marker. With one eye

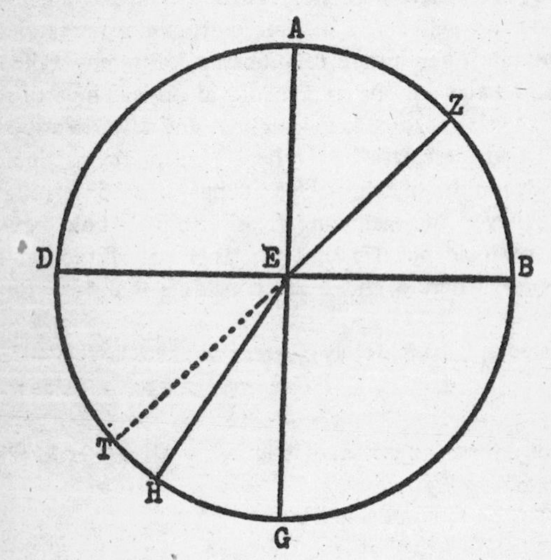

FIG. 11. The refraction of light.

take sightings until the markers at Z and E both appear on a straight line proceeding from the eye. At the same time move a small, thin rod along the arc, GD, of the opposite quadrant, which is under the water, until the extremity of the rod appears at the point of the arc which is on a prolongation of the line joining the points Z and E.

Now if we measure the arc between point G and the point H, at which the rod appears on the aforesaid line, we shall find that this arc, GH, will always be smaller than arc AZ. Furthermore, when we draw ZE and EH, angle AEZ will always be greater than angle GEH. But this is possible only if there is a bending, that is, if ray ZE is bent toward H, according to the amount by which one of the opposite angles exceeds the other.

If now, we place the eye along the perpendicular AE the visual ray will not be bent but will fall upon G, opposite A in the same straight line as AE.

In all other positions, however, as arc AZ is increased, arc GH is also increased, but the amount of bending will also be progressively greater.

When AZ is		GH will be about
When AZ is	10°,	8°
" "	20°,	15½°
" "	30°,	22½°
" "	40°,	29°
" "	50°,	35°
" "	60°,	40½°
" "	70°,	45½°
" "	80°,	50°

This is the method by which we have discovered the amount of refraction in the case of water. We have not found any perceptible difference in this respect between waters of different densities. (Cohen and Drabkin, *A Source Book in Greek Science*, pp. 274-275.[3])

[3] Quoted by permission of the publisher, the McGraw-Hill Book Company, Inc.; copyright, 1948.

Chapter 7

Greek Astronomy

As in some areas of physics, so in all of Greek astronomy from the fourth century B.C. we have a brilliant example of the fruitful application of geometrical techniques to scientific inquiry. Our knowledge of Greek astronomy before the fourth century B.C. is fragmentary. But we are told that the true nature of solar and lunar eclipses was discovered. Similarly, it was early asserted that the sun is the source of the moon's light and that the ecliptic, or sun's path through the fixed stars in the course of the year, is inclined to the celestial equator. There is a strong (if late) tradition that Pythagoras or the Pythagoreans believed the earth to be spherical in shape. For the most part it was thought that the earth resided in the center of a spherical universe, although certain Pythagoreans devised a system wherein the earth, a companion counter-earth, the sun, the moon, and the planets turned about a central fire, a system that was possibly the earliest of the geometrical systems devised by the Greeks to represent the complex movements of the heavenly bodies.

Crucial for the rise of mathematical astronomy toward the beginning of the fourth century B.C. was the problem of explaining the irregular movement of the planets as they wander through the fixed stars. This irregular activity of the planets could be reduced to a number of other movements:

(1) The planets rose and set nightly from east to west, as did all the heavenly bodies.

(2) Yet they also moved in the manner of the sun, from west to east through the fixed stars. Their courses could be plotted nightly through patterns or certain constellations of stars which were called the zodiac and which were bisected by the sun's apparent annual path through the sky. In fact, the west-to-east paths of the planets were found to be quite close to that of the sun. Each of the planets made the complete circling of the zodiac—i.e., returned to the same posi-

tion relative to the fixed stars of a given constellation of the zodiac—in a different period of time. Saturn was seen to take nearly thirty years; Jupiter, twelve years; Mars, two years and six months; Mercury, Venus, and the sun, one year.

(3) Now it was further noticed that, in addition to rising and setting nightly and circling the sky, the planets moved in still another fashion. In the course of their encirclement of the zodiac from west to east they were seen to slow down in their easterly movement until they appeared to stop in the sky, whereupon they seemed to reverse the direction of their

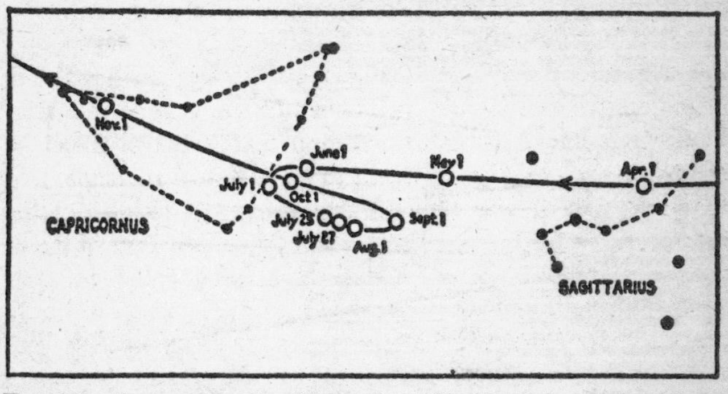

FIG. 12. Apparent path of Mars in 1939. Note the station about July 1, the retrogradation through July and August, and the second station about September 1. (Reproduced by permission of the publisher from R. H. Baker, *Astronomy*, 5th edition, copyright 1950, Van Nostrand Co., Inc.)

movement and proceed in a westerly direction for some time until they again stopped, reversing their direction once more to continue in their over-all easterly traversal of the zodiac. These points of apparent stopping are called stations, and the reverse westerly movement of the planet is called its retrograde movement.

If we plot these curious gyrations of the planetary movement against the constellations in the background, they will resemble a loop (see Fig. 12).

This phenomenon is readily explained on the basis of the Copernican system, which locates the sun at the center of the planetary system. For the planets, including the earth, revolve about the sun at various speeds, the planets closer to the sun moving with greater angular speed than those farther out. Since this is the case, the earth in traversing its orbit is continually overtaking the slower planets farther from the sun (the so-called superior planets: Mars, Jupiter, and Saturn) and at the same time is continually being overtaken by the planets closer to the sun (the inferior planets: Mercury and Venus). Now the planets are all revolving in the same easterly direction, but the earth's continuous change of position with respect to the other planets makes it appear to an observer on

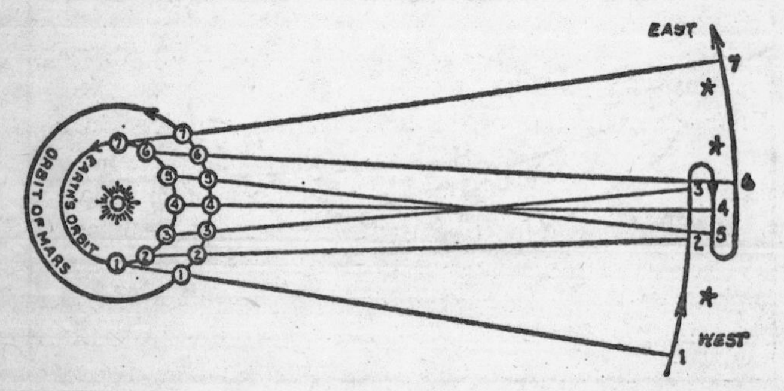

FIG. 13. Stations and retrogradation of Mars explained on the basis of Copernican theory. (Reproduced by permission of the publisher from R. H. Baker, *Astronomy*, 5th edition, copyright 1950, Van Nostrand Co., Inc.)

the earth who believes he is at rest that a given planet is reversing its movement. How this continuous change of position can produce the looping movement of stations and retrogradation is illustrated in Figure 13.

(4) One further aspect of planetary movement was observed and had to be accounted for by each of the systems devised to explain celestial movements; this was the motion of the planet in latitude. Here a word of explanation is required. The Greeks adopted the convention of using the zodiac as a kind of celestial map. The path of the sun, or ecliptic, which is the great circle running through the center of the circular zodiacal band with its twelve signs or distinctive star patterns,

was assumed as one reference line. A great circle perpendicular to the ecliptic and running through at the beginning of one of the signs, the ram (Aries), was used as the other reference line. The angular distance from this line of Aries along the ecliptic is called celestial longitude, and the angular distance along the line of Aries from the ecliptic is called celestial latitude. So far we have talked about the movements of the planets in longitude. These are the major movements, since the planets are seen to move quite close to the plane of the ecliptic. But in actuality the planes of the paths of the planets make small angles with the plane of the ecliptic. Hence the planets are continually changing their celestial latitude as well as longitude, and so any astronomical system has to account for this movement in latitude as well as for the major movements in longitude.

Now the objective of the mathematical astronomical systems devised by the Greeks was to account for the various classes of movement that we have outlined, in addition to certain less important movements. The over-all task of the astronomer was, as Simplicius has told us, "to invent hypotheses by which the phenomena will be saved (i.e., accounted for)." But, as Simplicius also pointed out, the astronomer in Greek antiquity borrowed his basic principles from the physicist, or, in actuality, the metaphysicist. Now the fundamental principle which the astronomer borrowed from the physicist at the very beginning of Greek mathematical astronomy was that celestial bodies as the result of their perfect nature move with a uniform, circular movement eternally. If the astronomer assumed the truth of this principle, then his problem became one of reducing the apparent irregularity of planetary movement to some combination of uniform, circular movements.

We are told by one author, Geminus, that "the Pythagoreans were the first to approach such questions, and they assumed that the motions of the sun, moon, and the planets are circular and uniform." But another author asserts that it was "Plato who set this problem for students of astronomy: By the assumption of what uniform and ordered motions can the apparent movements of the planets be accounted for?" It makes little difference whether it was Plato or the Pythagoreans, for the former seems to have borrowed heavily from the latter in such matters. This assumption of circular, uniform movement continued to be made for all astronomical systems devised until the time of Kepler, in the early seventeenth century. It was the basic assumption behind the system of Copernicus.

Author of the first mathematical system to reduce the irregular movements of celestial bodies to uniform, circular movements was the extraordinary, fertile mathematician Eudoxus, the student and associate of Plato. We have already celebrated his genius in mathematics as the perfecter of a more fecund theory of proportions and of the method of exhaustion.

For the basic astronomical problem Eudoxus devised a system of homocentric ("having the same center") or concentric spheres. It was outlined in a book entitled *On Speeds,* which is now lost. The geometrical details of the system have been reconstructed on the basis of two passages, one in Aristotle's *Metaphysics* and the other in Simplicius' commentary on Aristotle's *On the Heavens.* This reconstruction was accomplished in brilliant fashion by the Italian astronomer Schiaparelli, in 1875. It is Schiaparelli's reconstruction that is followed today and that we shall describe briefly, first noting Sir Thomas Heath's neat summary of it:

> Eudoxus adopted the view which prevailed from the earliest times to the time of Kepler, that circular motion was sufficient to account for the movements of all the heavenly bodies. With Eudoxus this circular motion took the form of the revolution of different spheres, each of which moves about a diameter as axis. All the spheres were concentric, the common centre being the centre of the earth; hence the name of "homocentric spheres" used in later times to describe the system. The spheres were of different sizes, one inside the other. Each planet was fixed at a point in the equator of the sphere which carried it, the sphere revolving at uniform speed about the diameter joining the

corresponding poles; that is, the planet revolved uniformly in a great circle of the sphere perpendicular to the axis of rotation. But one such circular motion was not enough; in order to explain the changes in the speed of the planets' motion, their stations and retrogradations, as well as their deviations in latitude, Eudoxus had to assume a number of such circular motions working on each planet and producing by their combination that single apparently irregular motion which can be deduced from mere observation. He accordingly held that the poles of the sphere which carries the planet are not fixed, but themselves move on a greater sphere concentric with the carrying sphere and moving about two different poles with a speed of its own. As even this was not sufficient to explain the phenomena, Eudoxus placed the poles of the second sphere on a third, which again was concentric with and larger than the first and second and moved about separate poles of its own, and with a speed peculiar to itself. For the planets yet a fourth sphere was required similarly related to the three others; for the sun and moon he found that, by a suitable choice of the positions of the poles and of speeds of rotation, he could make three spheres suffice. In the accounts of Aristotle and Simplicius the spheres are described in the reverse order, the sphere carrying the planet being the last. The spheres which move each planet Eudoxus made quite separate from those which move the others. One sphere sufficed of course to produce the daily rotation of the heavens [i.e., the fixed stars]. Thus with three spheres for the sun, three for the moon, four for each of the planets and one for the daily rotation, there were 27 spheres in all. It does not appear that Eudoxus speculated upon the causes of these rotational motions or the way in which they were transmitted from one sphere to another; nor did he inquire about the material of which they were made, their sizes and mutual distances. . . . It would appear that he did not give his spheres any substance or mechanical connection; the whole system was a purely geometrical hypothesis, or set of theoretical constructions calculated to represent the apparent paths of the planets and enable them to be computed. (T. L. Heath, *Aristarchus of Samos*, Oxford, 1913, pp. 195-196.[1])

Let us recapitulate briefly the theory as applied to the planets by referring to Figure 14. Sphere (1), the outermost sphere, rotates from east to west on its axis every twenty-four hours to account for the daily rising and setting of the planet.

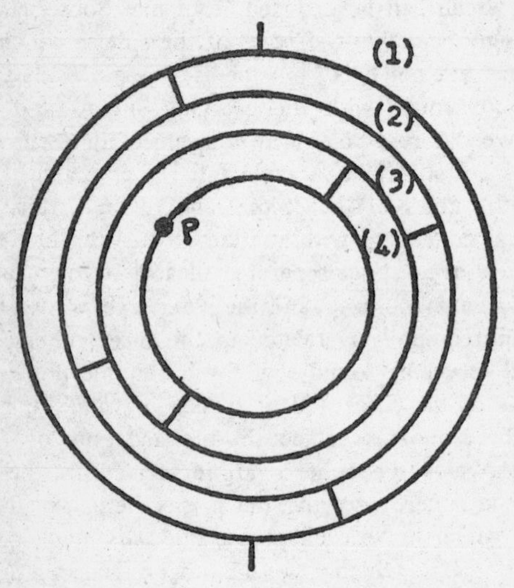

FIG. 14. A cross section of the four concentric spheres used by Eudoxus to explain planetary motion.

The poles of sphere (1) lie on a north-south axis. The rotation of sphere (2) accounts for the great encirclement of the planet from west to east through the zodiacal band; its axis is accordingly inclined to that of sphere (1) in about the same angle as the zodiacal band is inclined to the celestial equator (i.e., the equator of the first sphere). The third and fourth spheres rotate in equal times but in opposite directions. Together they account for the looping movement of the planet (i.e., for the stations and retrograde movement de-

[1] Quoted by permission of the publisher, the Clarendon Press, Oxford.

scribed above) and for some movement in latitude. The poles of the third sphere lie in the zodiacal band (i.e., in the equator of sphere (2)). The axis of the fourth sphere is inclined at an angle to the axis of the third sphere that varies for each planet, just as the speeds of spheres (3) and (4) vary for each planet. The planet (P) is carried on the equator of the fourth sphere. Now the combined movement of the third and fourth spheres will, as Simplicius notes, cause the planet to describe about the zodiac the curve called in Greek the hippopede, or "horse-fetter," which Schiaparelli shows to be

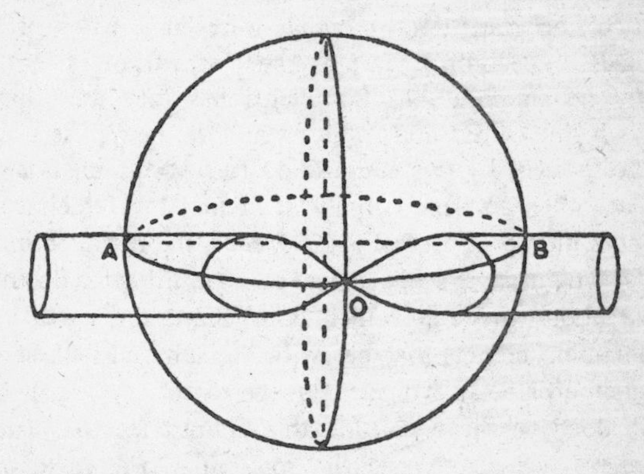

FIG. 15. The "horse-fetter" (spherical lemniscate) which the combined movements of Eudoxus' spheres (3) and (4) generate. Here described by the intersection of the surface of a cylinder and the surface of a sphere.

a spherical lemniscate (see Fig. 15). This curve bears a fair resemblance to the looping motion described by the planets.

That Eudoxus' theory had some currency we note from the fact that it was modified to include more spheres in order to take care of other observed irregularities; it was altered first by Callippus, an associate of Aristotle, and then by Aristotle, who, in an attempt to give the system some physical significance, arrived at a possible maximum of fifty-five spheres. The description of Eudoxus' system by Aristotle made it important for the medieval period. And even though it rather quickly passed out of favor in antiquity, it was revived from time to time under the authority of Aristotle by some Islamic and Western Latin authors. The fundamental observational evidence against the theory has already been noted in Chapter 2. It was the fact that the apparent size and brightness of the moon and of the planets varied, which seemed to indicate that they were at varying distances from the earth. And this observation ran counter to the hypothesis of concentric spheres.

The second important event in the development of planetary systems in Greek antiquity was the modification of the geocentric system suggested by another student of Plato, Heraclides of Pontus (*ca.* 388-310 B.C.). Heraclides is credited with two important doctrines that were later, when properly extended, to become a part of the Copernican system. (1) Initially he supposed, as Simplicius has told us, "that the earth is in the center and rotates while the heaven is at rest," and "he thought by this supposition to save the phenomena." (2) The other doctrine usually attributed to Heraclides was the assumption that Mercury and Venus, the so-called inferior planets, which always appear to be near the sun, do in fact revolve about the sun rather than about the earth, while the remaining planets together with the sun and its system of two satellite planets revolve about the earth.

Both doctrines have considerable interest for the historian of medieval science. When the rotation of the earth on its axis was seriously discussed in the fourteenth century at the University of Paris, it was Heraclides who was cited as the authority. The second doctrine was discussed by certain Latin authors in late antiquity and reappeared on occasion in medieval Latin writings.

We do not have to wait until early modern times for the extension of Heraclides' ideas, for it was accomplished by the Alexandrian astronomer Aristarchus (*fl. ca.* 281 B.C.), apparently another of the brilliant scholars at the Museum. To the rotation of the earth on its axis he added another movement, that of a revolution around the sun as a fixed center. Thus he brought to a halt both the apparent daily movement of celestial bodies and the annual movement of the sun through the fixed stars. It is to be assumed that Aristarchus also held that the other planets revolved about the sun in the manner of the earth. In short, Aristarchus appears to have arrived at the basic assumptions of the Copernican system almost two thousand years earlier than Copernicus. Aristarchus' original account of his system is lost, but Archimedes, who was only one generation removed from him, describes the system briefly in his *Sandreckoner*.

> Aristarchus of Samos brought out a book consisting of some hypotheses, in which the premises led to the result that the universe is many times greater than that now so-called. His hypotheses are that the fixed stars and the sun remain unmoved, that the earth revolves about the sun in the circumference of a circle, the sun lying in the middle of the orbit (*or*, in the circumference of a circle which lies in the midst of the course of [the Planets]), and that the sphere of the fixed stars, situated about the same center as the sun, is so great that the circle in which he supposes the earth to revolve bears such a proportion to the distance of the fixed stars as the centre of the sphere bears to its surface. (T. L. Heath, *Works of Archimedes*, Cambridge, Eng., 1897, pp. 221-222; modification by O. Neugebauer, *Isis*, Vol. 34 [1942], p. 6.[2])

The context of the whole passage makes it apparent that Aristarchus extended the distance to the fixed stars enormously in order to meet one of the main objections to his theory. This objection was that if the earth revolved about the sun, in different points in the orbit the angular distances between any two given fixed stars would change—i.e., would increase and then decrease in a regular fashion. In more technical language the objector would say that there should be parallactic displacement of the stars. A familiar analogy ought to make this clear to the reader. When one is riding on a train and looking at telephone poles ahead, these poles seem very close together; in fact, if we look far enough ahead the poles seem to be touching. Then, as we watch two poles and

[2] Quoted by permission of the publisher, the University Press, Cambridge, England.

come abreast of them, the distance between the poles grows to a maximum, and as we leave them behind they look as though they were coming together again. The objector would say that the same thing should happen if the earth were changing its position relative to the stellar telephone poles. But Aristarchus apparently met the objection by the assumption that the stars are so distant that the entire orbit is but a point in comparison, and hence the apparent displacement of stars would be so slight as to be unnoticeable. In fact, this displacement was not observed until the nineteenth century. The astronomers who came after Aristarchus shrank from this set of assumptions.

Plutarch tells us of a religious objection to the theory of Aristarchus, which, to the Hellenistic scientist, may not have been so important as the observational objection already noted. He notes that a certain Cleanthes (d. *ca.* 232 B.C.) had charged Aristarchus with impiety "for putting in motion the Hearth of the Universe, this being the effect of his attempt to save the phenomena by supposing the heaven to remain at rest and the earth to revolve in an oblique circle, while it rotates, at the same time, on its axis."

There are a number of other short passages which associate Aristarchus with a theory of the earth's motion. Although we have no idea of the extent to which Aristarchus developed his system, it would not appear unreasonable to call him the Copernicus of antiquity, or perhaps it would be preferable to call Copernicus the Aristarchus of modern times. Our only extant treatise by Aristarchus is one entitled *On the Sizes and Distances of the Sun and Moon,* which reveals a nice understanding of geometry and makes calculations with angles and the ratios of the sides of triangles that are equivalent to later trigonometric calculations. The results obtained were quite poor—e.g., that the distance of the sun from the earth is greater than eighteen times but less than twenty times the distance of the moon from the earth. But the reason for the poor results is that very poor observational data were assumed; the method employed (using similar triangles) is sound.

The attempt of Aristarchus to establish the distance and size of the sun and moon by geometrical methods was matched by their most successful application to the determination of the circumference of the earth by one of the ablest all-round scientists of antiquity, Eratosthenes of Cyrene. Eratosthenes, friend of Archimedes, librarian of the Museum at Alexandria, and founder of scientific cartography, was talented in so many fields that he was called *Pentathlos*—i.e., to indicate that he was an all-round figure in each of the scientific fields.

Eratosthenes' determination has been described in detail by a later Greek astronomer, Cleomedes; since it was this type of method that was used also in the Middle Ages, we shall report its highlights, referring to Figure 16. (1) It was assumed that the rays coming from the sun and striking various

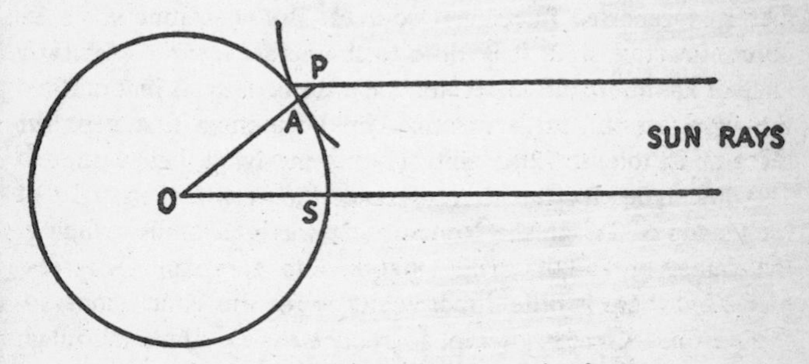

FIG. 16. Eratosthenes' determination.

parts of the earth are parallel. (2) It was observed (somewhat erroneously) that Alexandria (A) and Syene (S) were on the same meridian, 5000 stades apart. (3) When the sun was at the highest point in the sky at Syene on the day of the summer solstice, the pointer of a sundial cast no shadow. But at Alexandria it cast an arcal shadow on a bowl-shaped or hemispherical sundial equal to $\frac{1}{50} \cdot 360°$. (4) This shadow measures the angle at P (the angle of the sun at noon below zenith). But it also measures the angle at O, subtended by the arc AS, since the rays striking P and S are parallel, OS is the prolongation of the ray at S, and these parallel rays are cut by the straight line OP to form the equal angles at O and P. Hence the angle at O is $\frac{1}{50} \cdot 360°$. (5) The circumference is now immediately given by a simple proportion:

$$\frac{C}{360^\circ} = \frac{5000}{\frac{1}{50} \cdot 360^\circ}, \quad C = 250{,}000 \text{ (stades).}$$

He appears later to have arrived at a modified value of 252,000 stades as the result of further trials.

We are not sure which of the various stades ranging in value from 7½ to 10 stades to a Roman mile Eratosthenes used. Hence, we cannot be sure of his accuracy. A recent student of the problem has suggested the value of 10 stades to a mile, so that then the figure of Eratosthenes works out to 25,200 Roman miles (equal to 37,497 kilometers); while other students compute Eratosthenes' value as 39,690 kilometers, but in doing so they imply a stade which has an irrational value and is unreported in ancient sources. But the latter value is quite attractive since it is close to the actual mean circumference of about 40,120 kilometers. We are equally as ignorant of the value of the stade assumed by Posidonius (1st century B.C.) and Ptolemy (2nd century A.D.) in giving their value of 180,000 stades for the circumference. Some would say that a long stade of 7½ to the Roman mile was used, thus bringing the values of 252,000 and 180,000 into approximate agreement; but there is little direct evidence for this conclusion.

The fourth major attempt to reduce the apparent irregular movements of the celestial bodies to uniform, circular movements was the system of epicycles, or small circles having their centers on the circumferences of other circles. Its origins are obscure. One author has proposed that it was an invention of the Pythagoreans and that it is reflected in Plato's account. Others would say that it goes back to the fourth century B.C., to the efforts of Heraclides, and still others prefer, I think more correctly, to emphasize that it took form with the mathematical work of Apollonius of Perga (*fl.* end of third and beginning of second centuries B.C.) and his contemporaries. The epicyclic model was shown to be equivalent with that of eccentric circles and this equivalence was crucial for all succeeding astronomy, so that by the time of the second century after Christ, when the system received its most detailed exposition by Ptolemy, it included a mélange of eccentric and epicycle circles. Ptolemy's elaboration of the epicycle theory appeared in the *Mathematical Syntaxis,* or, as it was later called by the Islamic astronomers, the *Almagest.* This work is a superb blend of the mathematical and observational astronomy of antiquity. It was clearly the most influential single astronomical work until the time of the *De revolutionibus* of Copernicus (1543).

The nature of eccentric and epicycle circles is shown in Figure 17, which represents the movement of the sun according to both eccentric and epicycle circles. Thus, in the case of the eccentric system, E represents the earth and C is the center of the sun's circular path O. Hence the orbit or path of the sun is eccentric to the earth.

Now in the epicycle system, E is again the earth, but the sun, S, is assumed to be revolving in a small circle or epicycle about a mathematical point G, and G is assumed to be describ-

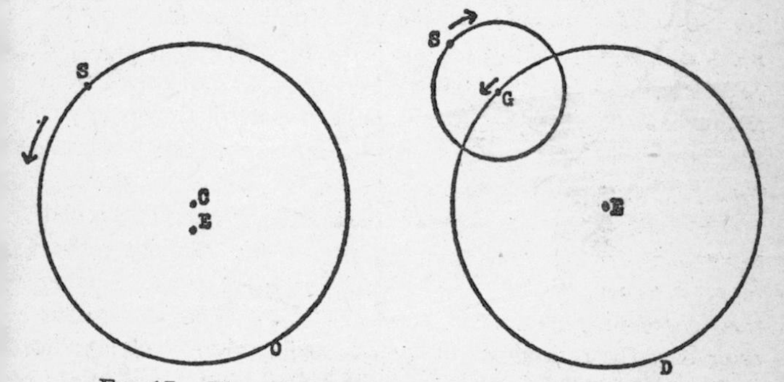

FIG. 17. The nature of eccentric and epicycle circles.

ing a circle D (called the deferent circle) about the earth as a center. The complete equivalence of these two ways of representing the movement of the sun was in all probability demonstrated by Apollonius (see Fig. 18). But it is suspected that Ptolemy was following the astronomer Hipparchus (second century B.C.), when, after noting the equivalence of the two systems for the sun's movement, he decided for the eccentric system, because it was the *simpler* of the two hypotheses, involving as it did only one circle. This observation of Ptolemy (or Hipparchus?) represents an interesting point of view of which we get occasional glimpses in antiquity: that it is the job of the astronomer (and the mathematical scientist in gen-

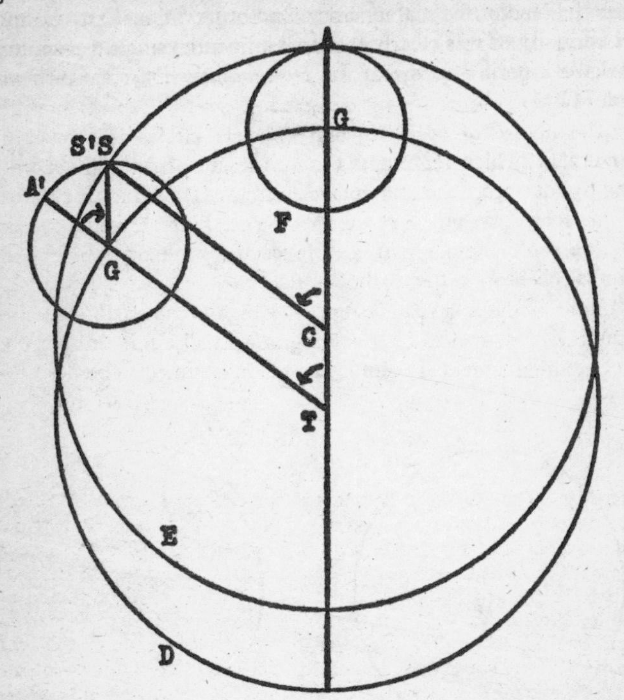

FIG. 18. The equivalence of epicycle and eccentric motions. The equivalence of the two models is easily proved (cf. Ptolemy's *Almagest,* Bk. III, Chap. 3). Paraphrasing the Ptolemaic proof and changing the letters, assume the earth *T, C* a point eccentric to *T* — which point *C* is the center of eccentric circle *E, G* the center of epicycle *F* — which point *G* moves uniformly and traverses circle *D* in the same period as point *S'* moving uniformly traverses epicycle *F* and as *S* moving uniformly traverses eccentric circle *E, AG* the radius of epicycle *F* — which radius *AG* is equal to eccentricity *CT*. To prove that *S* is always at the same position as *S'*, extend TG to *A'* and draw CS; then (1) *CSGT* is always a parallelogram, for $CT = SG$ and $CS = TG$ by construction; and (2) having assumed that arc *AS* on circle *E* is equal to arc *A'S'* on epicycle *F*, thus $\angle A'GS' = \angle ACS$; but (3) $\angle A'GS' = \angle A'GS$, for $\angle A'GS' = \angle ACS$ from (2), and $\angle ACS = \angle A'GS$ from (1); therefore *S* is in identical position with *S'*.

eral) to present the most economical mathematical system that will account for the phenomena, without regard necessarily for the real nature of the system. Economy, then, is a rule of rational procedure. Occasionally we find in antiquity and the Middle Ages the complementary but different idea that we adopt the most economical system because God or nature

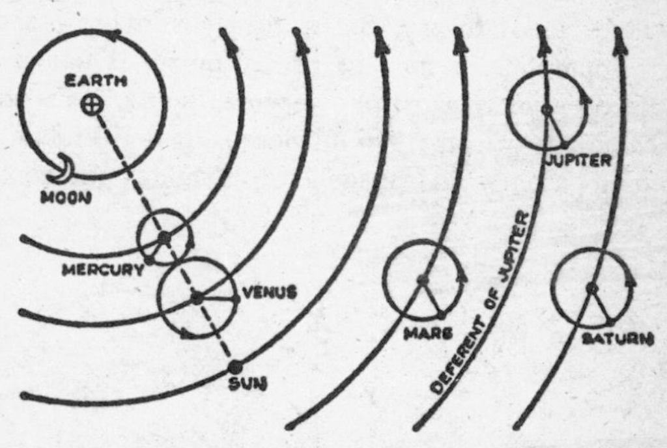

FIG. 19. The Ptolemaic "system" of planetary movements. It should be observed that this is an exceedingly simplified drawing in which the epicycle of the moon is omitted. Nor is any effort made here to reproduce the finer features such as eccentric centers of deferent circles or equant points (see Fig. 21). Furthermore, it should be noted that the Ptolemaic system in actuality is a collection of independent models rather than a truly connected system. Still this simplified diagram does reveal the fact that the sun's motion is somehow an important part of each model, for in the case of the interior planets the centers of the epicycle are always on the radius vector from the sun to the earth, while in the case of the exterior planets the radius from the center of the planet to the center of the epicycle is always parallel to the line from sun to earth. (Reproduced by permission of the publisher from R. H. Baker, *Astronomy*, 5th edition, copyright 1950, Van Nostrand Co., Inc.)

operates with the fewest possible causes—i.e., in the simplest way; hence economy of thought is necessary because of the economy of nature.

The details of how the eccentric and epicycle systems were merged can be ascertained by a careful study of O. Neugebauer's recent description of the kinematic models on which the tables of the Almagest were computed, at least those that allow us to compute for any given moment the longitudes of the sun, the moon, and the five planets. (O. Neugebauer, *Exact Science in Antiquity,* 2nd edition, Brown University Press, 1957, Appendix I.) It is important to realize that the Ptolemaic system represented the finest in Greek mathematical theory applied to astronomy. At the same time it contained reference to the observations of Hipparchus and his successors, which is not to say that it simply adopted the earlier values. Hipparchus, it may be briefly noted, is hailed as an important observational astronomer in antiquity. He is credited with a catalogue of some 850 or more stars and Ptolemy with new and apparently independent observations increased the

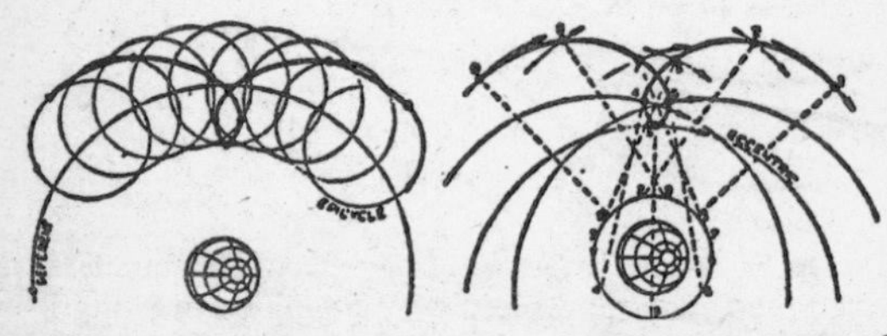

FIG. 20. How the epicycle movement of a planet accounts for the stations and retrogradations of the planet. (Reproduced by permission of the publisher from R. H. Baker, *Astronomy*, 5th edition, copyright 1950, Van Nostrand Co., Inc.)

number to about 1,000. By the careful comparison of his observations with earlier ones Hipparchus discovered the continual displacement of all of the stars relative to the equinoxes, i.e., that the longitudes of the fixed stars slowly increased. This phenomenon came to be called the "precession of the equinoxes." Also by a comparison of observations he concluded that the tropical year was not exactly 365¼ days long but was 1/300 of a day short of that figure. By using this new figure of the year's length, and the statement of Censorinus that Hipparchus determined a cycle of 304 years, we get a final figure of 111,035 days in the cycle. One purpose of such a cycle is to compute the mean month. This gives us as the length of the mean lunar month "29 days, 12 hours, 44 minutes, 2½ seconds, which is less than a second out in comparison with the present accepted figure of 29.530596 days!" (T. L. Heath, *Greek Astronomy*, London, 1932, p. liii). Hipparchus was able to achieve his success because of improved

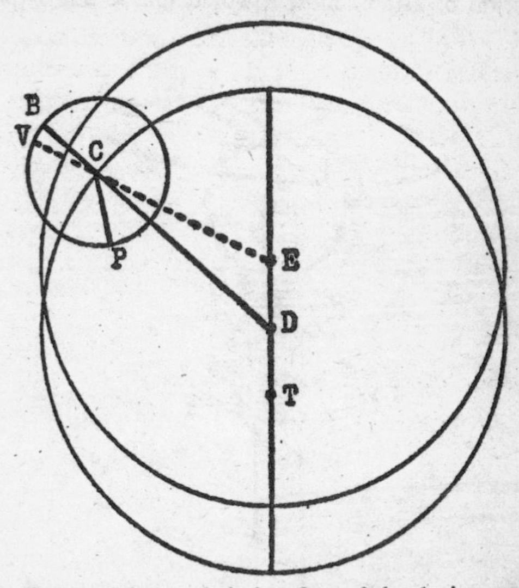

FIG. 21. Equant points and circles. One of the devices used in the Ptolemaic planetary models that modified the so-called Platonic rule of uniform motion was the "equant point" and its equant circle. Rather than simply having the center of the epicycle moving uniformly on the deferent circle (i.e. with respect to the center *D* of the deferent circle), the center *C* of the epicycle on which planet *P* moves is allowed to move uniformly with respect to another point *E*, the so-called equant point, which equant point is placed on the line of apsides on the opposite side of *D* from center of earth *T*. Or, to say it in another way, it is not line *DB* which sweeps out equal angles in equal times but rather line *EV*. *E* thus is thought of as the equalizing or "equant point" and any circle drawn with *E* as the center is an equant circle. In the case of Mercury the model was further complicated by the introduction of a moving equant point.

instruments (including an improved *dioptra*; see Fig. 22). Finally, we should note that Hipparchus and Ptolemy laid the foundations for trigonometry by computing the lengths of chords of various angles. Ptolemy gives us a table of the lengths of chords for each angle from ½° to 180°, at increments of ½° (see Appendix V).

Incidentally, in talking about the astronomical instruments of the period of Hipparchus, it would not be out of place to

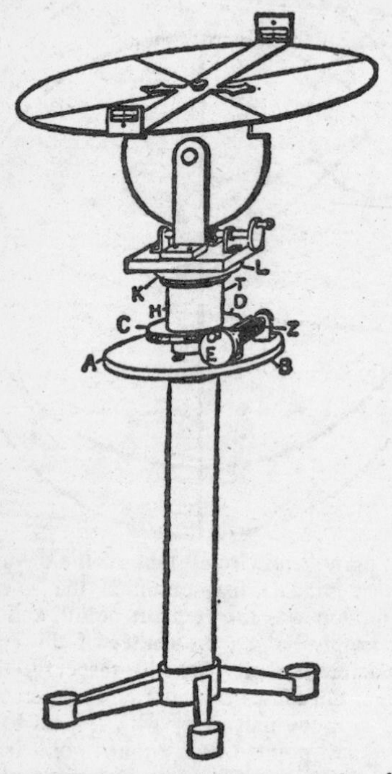

FIG. 22. A "dioptra," or sighting instrument, described by Hero of Alexandria. The gearing mechanism allows for changes in both azimuth and elevation. It was equipped with a hydraulic level to determine the horizontal plane.

mention the so-called Antikythera machine, an intricate clockwork and computing mechanism, recently reconstructed by Dr. Derek J. de Solla Price. Brought up from the bottom of the sea, where a shipwreck of about 65 B.C. deposited it, it survives as remarkable proof of the skill of ancient mechanics. Dr. Price's estimate of its significance ought to be quoted:

The original Antikythera mechanism must have borne remarkable resemblance to a good modern mechanical clock. It consisted of a wooden frame which supported metal plates front and back, each plate having quite complicated dials with pointers moving around them. The whole device was about as large as a thick folio encyclopedia volume. Inside the box formed by frame and plates was a mechanism of gear wheels, some twenty of them at least, arranged in a non-obvious way and including differential gears and a crown wheel, the whole lot being mounted on an internal bronze plate. A shaft ran into the box from the side, and when this was turned all the pointers moved over their dials at various speeds. The dial plates were protected by bronze doors hinged to them, and dials and doors carried the long inscriptions which described how the machine was to be operated.

It appears that this was indeed designed as a computing machine that could work out and exhibit the motions of the sun and moon and probably also the planets. Exactly how it did it is not clear, but the evidence thus far suggests that it was quite different from all other planetary models. It was not like the more familiar planetarium or orrery, which shows the planets moving around at their various speeds, but much more like a mechanization of the purely arithmetical Babylonian methods. One just read the dials in accordance with the instructions, and legends on the dials indicated which astronomical phenomena would be happening at any particular time. (Derek J. de Solla Price, *Science Since Babylon,* New Haven, 1961, pp. 40-41.)

To this point we have examined the principal achievements of the Greeks in evolving the general concept of a science, in developing the logical and mathematical tools—principally geometrical—needed for scientific investigation, and in realizing, however incompletely, something of the necessary relationship between theory and experience. We have seen how with these achievements in method the Greeks were able to build up a considerable body of scientific knowledge in medicine, in botany and zoology, in mathematics, and in physics and astronomy. In short, we have now examined the main elements of Greek science uncovered by an unending application of *historia*. Although some of the conclusions of Greek science rubbed off on the Romans, the concept of research epitomized in *historia* never caught on among them, as the documentation of our next chapter clearly demonstrates.

Chapter 8

Roman Science

I

In the second chapter we spoke of the basic chronology of Greek science and we labeled the last period, from about 100 B.C. to A.D. 600, the Greco-Roman period. But, although we spoke often in the succeeding chapters of writers who lived during this period, they were almost exclusively Greek or Greek-speaking. And this was no accident of selection, for Rome had little independent science, however great were her engineering achievements. Yet, unoriginal as Roman science was, its study is not without importance for the student of the history of science, since it was the Roman version of Greek science that was the main source of the scientific activity of the Latin West in the early Middle Ages.

In pursuing the main features of Roman science, we should keep in mind a few key observations:

(1) First, it should be realized that the Roman forte was practical, applied science. The word "applied" is used with some hesitancy, for more often than not the architect or engineer who covered the hills of Rome with mighty buildings or brought water within the city's limits over splendid aqueducts did not *consciously* apply scientific principles or methods to the solution of construction problems. But we do have a number of treatises in applied science that reveal a conscious attempt to stress some of the scientific theory behind and implicit in the engineering or technological activity. Such is the work on architecture of Vitruvius, that on aqueducts of Frontinus, the sundry works on agriculture, and a number of others which we shall treat briefly as we unfold the general chronological development of Roman scientific writing.

(2) With the emphasis on applied science came an apparent and deep-rooted disinterest in higher mathematics. It is extraordinary that not until the second century after Christ was a Greek mathematical work translated into Latin, and then it

was only a rather inferior beginner's manual. Even after the scientific translations became more common, in late antiquity, only Euclid among the great Hellenistic mathematicians—and perhaps not even all of Euclid—was rendered into Latin. It is no answer to say that the learned Roman knew Greek and felt no need to make the translations, for at least some scientific translations were made (see Chap. 10), but very few in mathematics.

(3) As students of Greek science without marked independent and original development, the Romans were attracted to the encyclopedic kind of presentation. This is demonstrated in the works of Varro, Vitruvius, Celsus, Pliny, and Seneca, as we shall see more clearly below. Interest in the independent scientific problem, so evident in the Hellenistic mathematical and physical treatises, is for the most part lacking in the Roman writings. The comprehensive manual, with emphasis on definition and description rather than on solution and demonstration, is the characteristic form of Roman science, just as it was the characteristic form of the scientific writings in the early Middle Ages which depended on the Roman science.

II

We can give only the most cursory glance to the marvelous engineering feats of the Romans and to their practical scientific achievements. Thus we can but mention the system of aqueducts that supplied Rome with millions of gallons of water each day, the remarkable sewage system of Rome, the fine harbors, the celebrated system of roads that linked Rome with all parts of its empire, and the military and public hospitals. Needless to say, these monuments attest the Roman genius for organization and engineering.

The practical nature of Roman scientific activity is illustrated also in the calendar reform initiated under Julius Caesar, which utilized Greek astronomical knowledge. The importance for administrative purposes of having a uniform calendar throughout the territory under Roman control was evident. And the importance for agricultural and other purposes of a calendar that corresponded with the seasons was also apparent. Replacing the lunar and other calendars then in use in various areas, the so-called Julian calendar assumed a solar year of 365 days and a leap year of 366 days every fourth year. This leap year was called the bissextile year, because of the addition of an extra sixth day, the *bissextus*, before the kalends, or first, of March. Hence the average length of the year in the Julian calendar was 365¼ days. But this year was slightly longer than the tropical year of the seasons, which runs from one vernal equinox to the next and is 365d, 5h, 48m, 46s. This made the Julian year too long by some three days in 400 years—i.e., the vernal equinox came a day earlier about every century and a third. In spite of this discrepancy, the Julian calendar persisted in Europe, not without some attempts at reform in the Middle Ages, until 1582; at that time Pope Gregory XIII ordered a modification, and the result is called the Gregorian calendar. Ten days were dropped out of the current year to bring the vernal equinox back to March 21. The new calendar compensated for the excessive length of the Julian year by making the century years regular years instead of leap years, except when those century years were divisible by 400 (e.g., 1700, 1800, and 1900 were regular years, whereas 1600 was a leap year, as will be 2000). As can be readily seen, this step removes three days in 400 years from the Julian calendar.

Also of a practical nature was the survey of the empire planned by Caesar but executed under Augustus. Augustus' son-in-law Agrippa (d. 12 B.C.) supervised the preparation of the survey and wrote a commentary in connection with it. The survey map then prepared was presumably the point of departure for a series of strategic maps prepared under the empire. These were route maps carrying accurate measurements of distances but probably not drawn on scientific cartographic principles, for if we can trust the medieval copy of one of such maps as being an accurate reproduction, there was a great distortion in the east-west direction.

III

The derivative character of Roman science is particularly evident in the fine poem *On the Nature of Things*, by Lucretius (from about 98 or 95 to 55 B.C.). Although not a work of science, it was to have great influence on scientists, not so much on medieval scientists—on whom its influence, if any, was indirect—as on the scientists of the early modern period, who in seeking to abandon Aristotelian philosophy turned to atomism. The principal source of the content of the poem was the teaching of the Greek atomist Epicurus. But the ethical aspects of Epicureanism are played down, and the result is a treatise on physical theory written in ringing poetic cadences; we must touch these difficult topics, Lucretius tells us, "with the Muses' delicious honey."

In the first book Lucretius initially attacks religion, "which has brought forth criminal and impious deeds" (I, 82-83). He denies creation by asserting the permanence of matter: "No thing is ever produced from nothing by divine power. . . . No thing can be created from nothing" (*nil posse creari de nilo*, I, 150-156). Matter is atomic in form. The atoms he posits as indivisible, permanent, invisible, varying in size, homogeneous as to character, and infinite in number. A void is assumed as a frame and a referent for motion. There are then two principles: atoms in motion and void. Time is an accident of bodies. It does not exist per se, "But rather from things (of nature) is derived the sense of what has been done in the past, of what is present with us, and further of what follows in the future" (I, 459-461). In the investigation of nature we must appeal to the senses; for "what can we find more certain than the senses themselves to mark the true and the false" (I, 699-700).

The second book is also important for the historian of science, for it deals with the kinetics of the atomic theory. The picture of the random movement, collisions, groupings, and separations of atoms to form the individual bodies that we know is a classic one (II, 80-130). The atoms wander through the void "carried on either by their own weight or by a chance blow from one or another." They often meet and collide while in rapid motion and so "they leap apart suddenly in different directions; and no wonder since as solids they are perfectly hard and nothing obstructs them from behind." "No rest is granted to the atoms throughout the profound void, but rather driven by incessant and varied motions, some after being pressed together then leap back with wide intervals, some again after the blow are tossed about within a narrow compass. And those being held in combination more closely condensed collide and leap back through tiny intervals. . . ." An analogical confirmation for the kinetic activity of the atoms exists in the activity of minute specks in the air revealed by the rays of the sun as they penetrate a dark room. These specks as they dance in the sunlight are as if "in everlasting conflict struggling, fighting, battling in troops without any pause, driven about with frequent meetings and partings; so that you may conjecture from this what it is for the primordial atoms to be ever tossed about in the great void."

Now Lucretius tells us that the atoms tend downward in the void (II, 184-251), although it would be interesting to know what meaning is given to the term "downward" in an infinite extent; however, it is guessed that he meant downward with respect to the earth. Although fire moves upward, it is only by compulsion, for it would naturally fall downward if not subjected to force by the air, which pushes it upward against its tendency to move downward. Without hindering bodies, all things would move downward in a void. Lucretius then tells us in a celebrated passage that things would fall with equal speed in a void. If things of different weight move with different speed through media because of different resistances, in a void which "cannot offer any support to anything anywhere . . . they must all be carried with equal speed, although they are not of equal weight." The reason why atoms can collide in a void in the first place is that they tend to swerve a bit. Otherwise there would have been no collisions and nature would have produced nothing, since all atoms would have fallen their separate parallel ways downward with equal speed. It is quite unclear what constitutes the weight of the falling atoms of Lucretius.

As we study Lucretius' account of the assembling and separating of the atoms to form the bodies that we sense, we see that the essential characteristics of the atoms (later one would speak of the primary qualities of the atoms) are their weight, their size, and their shape; but, although there is an infinite number of atoms, there is a finite number of sizes and shapes. Other qualities are reducible to these fundamental properties. The sensation of color is reducible to shape. Similarly, atoms are in themselves devoid of warmth and cold, barren of sound and moisture. "Nor do they throw off any odor of their own from their bodies." All these are derived qualities reducible to particular groupings of the root atoms possessing size, shape, and weight.

Lucretius' account in Book III of the material composition and mortality of the human soul we can pass over. We may also skip Book IV, which presents his somewhat complicated psychology and theory of sense perception, as they are not very significant to the history of science. On the other hand, his description in Book V of the origin of things is of some interest. Initially he here rejects the divinity of celestial bodies. Nor, he states, do they have any vital force. There is no home of the gods in the visible universe. This universe will some day pass away to be replaced by other combinations or universes. Our world is still young and developing, as is our study of it, and Lucretius asserts that he is one of the first students to describe the nature and the system of the world in Latin. He recounts how the world arose: not by design but by a chance assembly of atoms. In our universe the earth is at rest in the center. The cosmological details given by Lucretius are primitive and remind us of the opinions of the pre-Socratics. He recognizes the possibility of several worlds like our own, as did the earlier atomists.

In his discussion of the origin of things on the earth, Lucretius presents an evolutionary theory, which differs from the later Darwinian theory on some fundamental points. Both theories use the idea of natural selection or survival of the fittest. But in the theory of Lucretius the selection is made of chance atomic combinations that are better suited to survive; the combinations ill suited to survive die out. The entire concept of the variation and descent of species, fundamental to the Darwinian theory, is not present in the account of Lucretius. Thus in the Darwinian theory the selection is made of varieties of species more fit to survive, so that the species in its continuous history changes and evolves. But Lucretius describes the fate of the ill-fit as follows:

> But those to which nature gives no such (protective) qualities, so that they could neither live by themselves at their own will, nor give us some usefulness for which we might suffer them to feed under our protection and be safe, these certainly lay at the mercy of others for prey and profit, being all hampered by their own fateful chains, until nature brought that race to destruction. (Lucretius, *On the Nature of Things*, Book V, lines 871-877, Loeb Classical Library translation of W. H. D. Rouse, Cambridge, Mass., 1947.[1])

IV

Entirely different in character from the poem of Lucretius were the works of the learned Roman author Varro. Unfortunately Varro's influence on the Middle Ages was largely indirect, for his works disappeared at an early date. Varro is dated from 116 to 27 B.C. He was appointed curator of a future public library by Caesar in 47 B.C. By his seventy-eighth year he was reported to have edited some 490 books. All are lost but two, one on the Latin language and one on agriculture. But we have numerous fragments from the others. His most influential work so far as the history of science is concerned was his encyclopedia, *Nine Books on the Disciplines*. These disciplines included the later seven liberal arts of grammar, dialectics, rhetoric, geometry, arithmetic, astronomy, and music, plus medicine and architecture. We shall have occasion later to mention Varro's influence on such writers as Martianus Capella, who in turn were read and studied by medieval writers. Varro, then, is usually thought of as having set the original pattern for the liberal arts.

We are more fortunate in the case of Varro's junior contemporary Vitruvius, for his important *On Architecture*, written between 27 and 13 B.C., has come down to us intact. Like

[1] Quoted by permission of the present publisher and holder of the copyright, the Harvard University Press.

Lucretius and Varro before him, Vitruvius is heavily dependent on his Greek predecessors for his knowledge of the sciences. Although architecture is his main topic, Vitruvius covers a wide variety of scientific topics which he considers of importance for the architect. In the first chapter of the first book we are told that the science of architecture, itself consisting of practice and theory, is dependent on, or accompanied by, numerous other disciplines. The architect should be a man of letters, a draftsman, a skilled geometer familiar with scientific investigations, a student of philosophy with a knowledge of music and of medicine, a man who knows law and has an understanding of astronomy. He must, further, be acquainted with optics. In giving the detailed reasons why the architect should be familiar with other sciences Vitruvius gives some sound arguments, as well as some very labored ones. Thus he tells us that "philosophy explains the nature of things, which in Greek is called *physiologia*, a subject which it is necessary to have studied carefully because it presents many different natural problems, as for example, in the case of water conduction. . . ." Natural air pockets are produced in the aqueduct courses, causing difficulties "which cannot be remedied unless one has learnt from philosophy the principles of natural phenomena." And so the architect, to meet such situations, should read the works of Ctesibius (a mechanician of Alexandria dated about 283-247 B.C.?) and Archimedes. Furthermore a knowledge of astronomy, he says, is necessary for the study of clocks. He then pays tribute to Greek mathematicians and mechanicians, and he lists a few of them.

Vitruvius clearly practiced what he preached, for he shows a surprisingly good knowledge of Greek science. Among the observations of interest to the historian of science we may note his opinions on the importance for man of the discovery of fire and human discourse, as well as on the importance of man's life in common and the advantage gained by the use of his hands:

> Therefore, because of the discovery of fire, there arose at the beginning, concourse among men, deliberation and

a life in common. Many came together in one place, having from nature this boon beyond other animals, that they should walk not with their head down, but upright, and should look upon the magnificence of the world and of the stars. They also easily handled with their hands and fingers whatever they wished. Hence after thus meeting together, they began, some to make shelters of leaves, some to dig caves under the hills, some to make of mud and wattles places for shelter. . . . (Vitruvius, *On Architecture*, Book II, Chap. I., Loeb Classical Library translation of F. Granger, London, 1931.[2])

Meteorologists will find interesting Vitruvius' description of wind currents (I, 6) and, particularly, of the wind tower built at Athens by Andronicus of Cyrrha. Vitruvius describes the tower as an octangular marble structure and on the several sides of the octagon, there were "representations of the winds carved opposite their several currents. And above that tower he (Andronicus) caused to be made a marble upright and above it he placed a bronze Triton holding a rod in his right hand. He so contrived it that it was driven around by the wind and always faced the current of air, and held the rod as an indicator above the representation of the wind blowing."

The ninth book is devoted to astronomy; although it is quite elementary, there are some passages worthy of note. In the first chapter Vitruvius alludes to the doctrine of the revolution of Mercury and Venus about the sun, without mentioning the name of its author, Heraclides (see Chap. 7): "The stars of Mercury and Venus make their retrograde motions and retardations about the rays of the sun, forming by their courses a wreath or crown about the sun itself as center" (T. L. Heath, *Aristarchus of Samos*, Oxford, 1913, p. 255). Another section (IX, 8) tells us of the construction of water clocks attributed to Ctesibius. According to Vitruvius, Ctesibius, who was the son of a barber, discovered the nature of wind pressure and the principles of pneumatics. We get the impression that Vitruvius believes him to be the father of the Greek mechani-

[2] Quoted by permission of the present publisher and holder of the copyright, the Harvard University Press.

cians whose works we have described in Chapter 6. The clocks described by Vitruvius are of considerable ingenuity.

Book X of *On Architecture* he devotes to mechanics. A machine he defines as "a continuous material system having the power to move weights." He classifies machines as (1) ladders, (2) pneumatic machines, and (3) traction machines depending on equilibrium. The machines described resemble those found in the works of Philo and Hero and are often given in great detail. And should anyone think that the machines devised were used only as toys, as the materialists often say, let him read Chapter 2 (section II), where there is described an ingenious device for bringing the shafts to be used as columns from the quarries to the Temple of Diana at Ephesus, and let him also read Chapters 4-7, which detail devices to be utilized for the raising of water. Included in the latter discussion is a pump attributed to Ctesibius. We know that water pumps of this kind were utilized in the water system of the Romans, for the remains of such a pump from Bolsena in Etruria is now extant at the British Museum. Note should be made also of the hydraulic organ described by Vitruvius (X, 8) and Hero (*Pneumatics*, I, 42). It was supposedly invented by Ctesibius and used in late antiquity and the early Middle Ages by the Christians. Of similar interest is Vitruvius' description of a hodometer to measure the distance traveled by a four-wheeled carriage or by a ship (X, 9; cf. Hero, *Dioptra*, 34).

The principal manuscripts of *On Architecture* are of the early medieval period, and the best manuscript comes from the eighth century. Hence we must conclude that Vitruvius was not without some influence in the early medieval period. Those fortunate enough to have a copy of Vitruvius' work at hand were in contact with a writer who had some real understanding of Greek science.

V

Even more closely dependent on Greek sources than Vitruvius was the Roman medical writer Celsus, who composed a treatise called *On Medicine*. Little is known about Celsus. He appears to have lived in Narbonne and to have been born about 25 B.C. It is quite possible that this treatise was the second part of a scientific encyclopedia, which included books on agriculture, military science, rhetoric, philosophy, and jurisprudence. It is a moot question whether Celsus was a physician or merely a capable summarizer. It is further mooted whether the medical work represented an original editing and paraphrasing of Greek medical sources or whether it was a direct translation of a Greek original. Whoever composed the treatise, it demonstrates a firm grasp of Greek medicine as it existed about the first century after Christ. The introductory section, from which we have quoted in Chapter 4, gives a summary of the history of Greek medicine. Its value to the historian of science lies in its references to the principles of method followed by Greek medical practitioners. The author follows neither the Dogmatists nor the Empiricists exclusively, but seems rather to assume a middle course. One of his more recent editors has described the anatomical knowledge of Celsus as fair. Four of the chief manuscripts of *On Medicine* date from about the ninth and tenth centuries and show that the work had some currency in the early Middle Ages.

We have already spoken of the progress of the Romans in the development of military and public hospitals and of their remarkable sewage system in Rome, the so-called Cloaca Maxima, which emptied into the Tiber. And reference has likewise been made to the Roman aqueduct system. A description of the latter has come down to us from Frontinus, who flourished from A.D. 35 to 103. Entitled *On the Waters of the City of Rome*, it has scientific interest because of its description of the volumetric measures used in computing the capacity of the aqueducts. This part constitutes a nice dissertation on measures. It demonstrates an appreciation of the facts that the pressure of water flow increases with the height or head of the water system and that the volume of water delivered is reduced as the length of the conduit is increased, "owing to the resistance of the conduit."

Although the Romans never achieved anything like the independent theoretical investigation of natural history of Aristotle and Theophrastus, they were keenly interested in the subject, particularly in its practical aspects. Thus we have a number of good agricultural treatises written in Latin, as well as an excellent treatise on veterinary science by Renatus (A.D. 383-450). On the less practical side, encyclopedic works such as that of Pliny contain much of interest to the nature lover.

VI

In our discussion of Celsus and Frontinus we have pushed well into the first century of our era. And in this century we are brought face to face with the celebrated authors Seneca and Pliny. These men were neither theoretical scientists nor specialists in one of the arts or crafts. And yet they cannot be left out of any discussion of the history of science, because their influence on medieval writers was marked.

We cannot detail here the life of Seneca, but we should note that he was born toward the beginning of our era (estimates run from 8 to 1 B.C.), in Cordova, Spain. Although he appears to have studied in his youth many of the numerous sects and philosophical movements of his day, he reflects the Stoic philosophy more than that of any other school. Seneca spent many years in political positions, playing the game of palace intrigue which finally, after Nero's dissatisfaction with him, resulted in his enforced suicide.

The only one of Seneca's many writings of interest to the historian of science is his *Natural Questions*, written soon after A.D. 63. The questions treated covered varied topics in astronomy, meteorology, physics, and physical geography. In the form that the work has come down to us it is a potpourri, badly organized. A distinctive feature of the work is that the author frequently draws morals for man's conduct from his exposition of natural questions. Thus his most interesting description of an earthquake in A.D. 63 is followed by an essay on meeting death stoically.

Seneca's sources were primarily Greek and included Aristotle's *Meteorology*, his *Book of the Heavens*, and possibly the pseudo-Aristotelian *De mundo* (*On the Universe*); Posidonius' *Meteorologica*; the writings of Theophrastus; and the works of a number of less important writers. Just as he appreciated the Greek sources, so he understood the essentially progressive nature of the development of science. This he tells us in a moving passage:

> Why should we be surprised, then, that comets, so rare a sight in the universe, are not embraced under definite laws . . . seeing that their return is at long intervals? It is not yet fifteen hundred years since Greece "counted the number of the stars and named them every one." And there are many nations at the present hour who merely know the face of the sky and do not yet understand why the moon is obscured in an eclipse. . . . The day will yet come when the progress of research through the long ages will reveal to sight the mysteries of nature that are now concealed. (Seneca, *Natural Questions*, Book VII, Chap. 25, translation of J. Clarke, London, 1910.)

Seneca's style is loose and rambling, and the evidence of independent research is generally lacking. Among the topics he treats are meteors, halos, and rainbows; the nature of air, thunder, and lightning; the forms of water, rivers, and seas; snow, hail, and rain; the winds and atmospheric movements; earthquakes; and comets. In drawing a moral on the movements of comets and the various explanations thereof, how like the Christian authors such as St. Basil he sounds:

> Nor is it for man that God has made all things. How small a portion of His mighty work is entrusted to us? But He who directs them all, who established and laid the foundations of all this world, who has clothed Himself with creation, as is the greater part of His work. . . . (*Natural Questions*, Book VII, Chap. 30.)

Pliny, Seneca's contemporary, is in some ways analogous to the Stoic philosopher. Born in A.D. 23 or 24, he died at the spectacular eruption of Vesuvius in 79, a martyr to scientific curiosity. Like Seneca his life was one of public service, and like the Cordovan he drew morals from nature, but in our opinion without the hypocrisy of Seneca. He entered military service and served the emperors Vespasian and Titus in various capacities.

The work that brings Pliny into the annals of science is his voluminous *Natural History*, clearly one of the most important books in the Latin tongue. Some thirty-seven books of the *Natural History* remain as a storehouse of antique learning. As he tells us in his dedication, he has included "20,000 topics all worthy of attention . . . gained by the perusal of about 2000 volumes . . . of one hundred select authors." (Modern students of Pliny believe they have found some 400 to 500 authors used.) One of the striking things about Pliny's technique is his scrupulous citation by name of the authorities he has employed. He tells us he thinks it courteous and modest to acknowledge sources from which he has derived assistance. Although on occasion he attacks Greek authors, his work is either directly or indirectly dependent upon them.

Since the *Natural History* is a compilation, its accuracy and value vary from book to book in accordance with the sources Pliny utilizes, as is frequently true of encyclopedias. Little of the material is the result of his own scientific research. But he does show critical acumen in places, and some of the stories cited to prove his credulity are often simply evidences of his attraction to curiosities, showing rather the antiquarian than the devotee of the fabulous and marvelous. Perhaps the key to his attitude is revealed by the statement of his nephew Pliny the Younger to the effect that his uncle thought that no book was so bad that some good might not be gotten from it. Lynn Thorndike has pointed out that, although Pliny condemns magic, he often seems unable to distinguish between magical and natural phenomena, an inability shared by many of the writers of his and a later day, and perhaps symptomatic of the second-hand, encyclopedic type of learning.

The general arrangement of the materials of the thirty-seven books of the *Natural History* is worth noting. Book I is dedicatory and contains a table of contents. The second book summarizes astronomical knowledge. Then follow four books on geography and a book on man and his inventions. Books VIII-XI are zoological, describing various animals with real and fabulous properties. Books XII- XIX are botanical in content, and they are followed by eight books that deal with medicines derived from plants. Books XXVIII-XXXI deal with medicines compounded from man and animals. Book XXXII treats of the properties of aquatic animals, and the last books are concerned with metallurgy, plants, and gems.

Among the most interesting of Pliny's chapters is the one (II, 112) on the dimensions of the earth. Here he rather grudgingly reports the value of 252,000 stades found by Eratosthenes, which value he equates to 31,500 Roman miles. He notes that the value is presented with such subtle arguments that we must give assent to them. He gives at the same time a larger value which he attributes to Hipparchus. At the same time he cannot resist (nor can we) giving a most remarkable determination of one Dionysidorus, a determination that Pliny, no doubt with his tongue in his cheek, characterizes as "less worthy of confidence."

> He was a native of Melos, and was celebrated for his knowledge of geometry; he died of old age in his native country. His female relations, who inherited his property, attended his funeral, and when they had for several successive days performed the usual rites, they are said to have found in his tomb an epistle written in his own name to those left above; it stated that he had descended from his tomb to the lowest part of the earth, and that it was a distance of 42,000 stadia. There were not wanting certain geometricians, who interpreted this epistle as if it had been sent from the middle of the globe, the point which is at the greatest distance from the surface, and which must necessarily be the center of the sphere. Hence the estimate has been made that it is 252,000 stades in circumference. (Pliny, *Natural History,* Book II, Chap. 112, translation of J. Bostock and H. T. Riley, London, 1855.)

The more we learn of early medieval science the more we realize that Pliny, at least in excerpt form, was quite widely known in the early Middle Ages, from the eighth century onward. And before that time he was known through St. Isidore and others. The work of one of his plagiarizers, Solinus (who wrote some unknown time after Pliny), was also known, particularly to St. Isidore, whose encyclopedia, which will be discussed later, was a most widely read work. Also current was a collection of medical excerpts entitled the *Medicine of Pliny*. And, as we shall see in a later chapter, his statements on astronomy were excerpted and became part of a computistical corpus that attained wide popularity from about the eighth century.

The story of Roman science does not, of course, stop with the figures of the first century. But we should like to postpone presenting the Roman science of late antiquity until a later chapter, so that we may examine the transition to medieval science more closely. For there is a clearly discernible set of ties between the Roman science here presented and the Latin science of the early Middle Ages. But before we examine these ties, something should be said on the effect on science and on learning in general of the spiritual forces loose in the empire.

PART TWO

Science in Late Antiquity

Chapter 9

Science and Spiritual Forces in Late Antiquity

I

IN THE first part of this volume we have traced the rise of Greek science and its development to maturity. We must now investigate the fate of that science in late antiquity. We must see what modifications and reorganizations the Greek corpus of science and philosophy underwent in late antiquity, in preparation for its long journey as it passed successively into Syriac, Arabic, and Latin.

It has already been suggested that Greek science did not decline radically in late antiquity, at least to the point where it was no longer worked over and studied. We have said that it leveled off. This leveling off was undoubtedly tied up with complicated social and political changes brought about in the Mediterranean area by the rise and spread of Roman power. But it would have taken a fine eye in the first two or even three centuries of the Christian era to detect any decline in Greek science or the Greek rational spirit by an examination alone of the works of the best scientists.

Let us suppose that our investigator examined the extraordinary activity of Ptolemy in the second century. In the *Almagest,* he would, as we have already noted, find one of the highest products of Greek astronomy, the author of which is clearly a peer of the best that the Hellenistic period had to offer in astronomy. Or, if he were fortunate enough to read Ptolemy's other works, he would see other scientific topics being treated with the same fine critical spirit in conjunction with mathematical and observational techniques. After reading the *Optics,* usually attributed to Ptolemy, our investigator would conclude that this was certainly the best of the numerous optical treatises that came out of Hellenistic and post-Hellenistic times. Nowhere else would he find so nice a juncture of mathematics and experiment, even though the mathematics is of an elementary nature. On turning to Ptolemy's *Geography*, he would see a work that extends the Hellenistic beginnings in mathematical-scientific cartography. Furthermore, he would find in the *Planisphere* some of the fundamentals of stereographic projection. He might begin to wonder if he had not found some evidence of a nonscientific approach as he read Ptolemy's astrological work, the *Tetrabiblos*. But even here the critical spirit is present.

On pursuing his inquiry into the scientific works of the second century, our investigator would examine, if he had time, the 153 books of Galen. He would find among them such works as *On the Use of Parts* and *On the Natural Faculties*, which contain the best in Greek medical and biological research; and if he found errors and petty argumentation, he could find them, too, in the writings of Greek scientists of an earlier period. But we ask him particularly to read the passage quoted in Chapter 4, Section I (and numerous others like it), which shows Galen as an imaginative proponent of an experimental procedure.

Now it might be said with some justification by our investigator that some of the other scientists, particularly a number of the mathematicians of the first centuries of our era, were mediocre products of Greek science. Thus Nicomachus, who flourished about A. D. 100 and who seems to have been a Pythagorean, wrote an inferior *Introduction to Arithmetic*. But it should be remembered that we tend to compare Nicomachus with the very best men, such as Euclid or Archimedes. Of course he does not come up to their genius; nor does he compare with Ptolemy, or Diophantus, or Pappus—men close to his own time. In fact, Nicomachus is important to the historian of science mainly because the Roman statesman and philosopher Boethius, on the very edge of the Middle Ages, in the sixth century, wrote a manual of arithmetic based on Nicomachus, and thus indirectly established him as the chief authority on arithmetic, whose work was available in Latin in the early Middle Ages.

If Nicomachus was not in the same class with the great geometers of the Hellenistic period, his work nonetheless bears the imprint of Greek rationalism. The same can be said for two authors of the first century who wrote on the subject of spherical geometry, Theodosius and Menelaus. Both were compilers (but so was Euclid). Yet the *Spherics* of Menelaus was an advanced and perhaps even original work. It included a treatment of spherical triangles that was more comprehensive than Euclid's presentation of plane triangles. And its third book is one of the landmarks in the early history of trigonometry. Because of their importance for astronomy, the works of both Theodosius and Menelaus were translated in the high Middle Ages and became part of the medieval mathematical-astronomical corpus.

We have already suggested that Galen and Ptolemy were not the only authors of the early Christian era who represented Greek science at its highest level. Hero of Alexandria also belongs to that select group. We have already discussed his *Mechanics* as being the culminating effort of mechanics in late antiquity (see Chap. 6) and as containing both theoretical and applied mechanics. His writings, particularly the *Metrica*, which included many formulae, and his commentary on Euclid's *Elements* (of which parts remain in Arabic) reveal him as an excellent mathematician.

It is generally agreed that one of the most powerful and skilful of the Greek mathematicians was Diophantus of Alexandria (*fl. ca.* A.D. 250). A kind of geometrical algebra had existed from the earliest days of Greek mathematics, and probably from the time of Euclid it had even been used to find numerical answers; Diophantus in his *Arithmetic* greatly extended algebraic analysis beyond the point of the earlier Greek mathematicians. In addition, his use of algebraic notation was, so far as we know, original. In the development of his algebra he may have been dependent in some fashion on the well-developed Babylonian algebra or even on the less advanced algebra of Egypt. But neither of these early algebras used symbols. We shall go into the question of algebra a little more extensively in a later volume, when we discuss its further maturation in Islam; but we can note here in passing that Diophantus used a sign for the unknown in an equation (something on the order of the modern usage of x for that quantity); similarly, he had special signs for powers of the unknown, such as x^2, x^3, etc. But since he had only one sign for the unknown instead of several, such as x, y, z, whenever he dealt with more than one unknown he had to use great skill to avoid confusion. Diophantus also used a symbol for subtraction, a minus sign. Thus he used symbols on the one hand for an unknown quantity and on the other for operations. Since the establishment of symbols was one of the real stimuli to the rapid development of algebraic analysis in early modern times, it is indeed unfortunate that Diophantus' initial efforts produced little response among medieval algebraists. Diophantus easily solved quadratic equations (involving the unknown squared) and in one special case a cubic equation. His name is still connected with the solution of what are called indeterminate equations. There can be little doubt that Greek mathematical genius was still burning brightly at the time of Diophantus.

Scarcely less gifted was the mathematician and mechanist Pappus of Alexandria, a junior contemporary of Diophantus, who flourished in the period 284-305. Pappus apparently set as his goal "the re-establishment of geometry on its former high plane of achievement" (T. L. Heath). Pappus' activity was prodigious. Among his works, extant only in part, were a commentary on the *Elements* of Euclid and commentaries on the *Mathematical Syntaxis* and *Planisphere* of Ptolemy. Probably his greatest work was his enormous *Mathematical Collection,* which covered the whole field of geometry. Intended as a handbook to be used in conjunction with the great treatises of geometry, it nevertheless gave many alternate and independent proofs. It is a gold mine of historical information, mentioning and discussing most of the famous geometers of the past. Sir Thomas Heath, one of the ablest historians of Greek mathematics, has confirmed that "the whole work shows on the part of the author a thorough grasp of all the subjects treated, independence of judgment, mastery of technique. . . ; in short Pappus stands out as an accomplished and versatile mathematician, a worthy representative of the classical Greek geometry."

II

It should be unnecessary to extend further the evidence that, although Greek science was not everywhere soaring to its greatest heights, it flourished deep into the critical period of late antiquity as a monument to the persistence of the Greek rational spirit.

Yet if our investigator were to look beyond the more obviously scientific works, he would find an impressive amount of evidence of the effects of spiritual forces that were predominant in the Empire and that were to affect noticeably the pursuit of natural knowledge. Now of course there were always such currents alive in the Greek and Hellenistic world. But by the first centuries of the Christian era, these forces were helping to produce profound social and intellectual changes.

We shall not attempt here to assay critically the spread of spiritual and religious movements. No doubt the merging of the Greek and Near Eastern cultures in the Hellenistic period, a merger confirmed by the establishment of Roman hegemony over the entire area, provided an opportunity for that spread. It would be incredible that the Near East should be Hellenized (however lightly) without exerting some counterinfluence, first on the ruling Greeks and Macedonians and later on the Romans. Numerous sects that were to multiply in the first centuries of the Christian era were strongly indicative or reminiscent of a Near Eastern origin. While the intermixture of the cultures offered the opportunity for the spread of the spiritual movements, no doubt fundamental political and economic causes favored their rapid growth in the Empire. One historian of Roman religion, T. R. Glover, gives us a hint of this when he speaks of "a rule that robbed men of every liberating interest in life and left society politically, intellectually, and morally sterile and empty." Not only did "a flood of vulgar superstition" attempt to fill this moral vacuum, but also religious movements of higher character.

Let us take, for example, the cult of Isis, which spread from Egypt throughout the Mediterranean. In its early stages it appears to have been of an excessively superstitious nature, if we can believe the Roman writers who scornfully labeled it a sanctuary of superstition and loose women. But by the end of the second century it was possessed of an organized clergy, prayer books, vestments, secret initiations, a form of baptism, and fasting. Isis was the all-merciful mother "who identified all the Divinities with herself" and "who was the power underlying all nature."

Still other cults and offshoots were to influence the intellectual activity of the Empire. We can mention, for example, the Gnostic sect, for it appears to have had some influence on the mystical side of alchemy. Gnosticism has been considered until very recently a rather debased offshoot of Christianity. This opinion resulted largely from our reliance on the opponents of Gnosticism for our judgments. But the discovery recently of a block of some 1000 papyri containing in Coptic (the language of Egypt) the religious books of Gnosticism portends a change in the evaluation of Gnosticism. Preliminary examination of this material has already shown Gnosticism as a vital, perhaps independent, religious movement with a significant sacred literature of its own.

But the Christian church soon outdistanced its rivals, both as to membership and as to influence. Its rapid growth in the third century was met by a more forceful reaction from the Empire. Where persecutions before had been occasionally spectacular but really inconsequential, from 250 onward political persecution on a large scale became more frequent until 313, when Christianity won its major victory and was accepted on equal footing with the other religions of the Empire. In fact, its influence soon greatly exceeded that of the other religions, and within a century after its acceptance it was, for all intents and purposes, the state religion. What is important for the historian of science is that from about the year 200 the Church began to attract in fairly significant numbers men who might have gone into philosophy or science but now undertook the writing of Christian apologetic and dogmatic literature. And, as we shall see in the next chapter, when these early fathers of the Church concerned themselves with scientific matters their opinions were naturally colored by their moral and religious predispositions.

III

One of the most interesting evidences of the influence of the spiritual movements on intellectual activity was the so-called Hermetic literature. This was a collection of works attributed to the god Hermes Trismegistus ("the thrice-great Hermes") or to his Egyptian counterpart, Thoth. It included treatises on many subjects, including alchemy, astrology, astronomy, physics, embryology, botany, medicine, and numerous topics of a purely magical, mystical, or religious character which are not of concern to us.

Why were these writings attributed to Hermes or Thoth? From the earliest times, the Egyptian god Thoth was designated as the secretary of Osiris and then as the inventor of writing. Later he became the patron and discoverer of the sciences. After Greek mythology fused with Egyptian in the Hellenistic period, Thoth or Hermes began to be designated as the author of various treatises of a religious and quasi-scientific character.

Dating of the origin of these writings is elusive. Although some references to such treatises apparently go back at least to the first century B.C., more clear-cut references appear in the following century. By the second century a number of treatises existed, according to contemporary references.

The motivation behind attribution of the writings to a god is obvious—to give the treatises in question the authority of sanctity and age. As revealed literature they pretended to be the word of Hermes and therefore to be beyond contradiction. The spread of the revelation technique to topics of scientific interest constituted a profound alteration in scientific literature. In a sense it was the Greek miracle in reverse. In the Hermetic corpus the appeal to supernatural causation and to the continuous and direct influence of supernatural on natural phenomena was made on a wide scale. Divine revelation as the source of truth, even of scientific matters, also became characteristic of Christian writing, and it persisted into the Middle

Ages even when it was played down in favor of empirical evidence.

In spite of the revealed character of the Hermetic writings, their most recent and authoritative student, A. M. Festugière, has come to the conclusion that they do not represent the work of a single religious order; they were not the "bible" of a sect. Rather, Hermeticism was one of the forms that Hellenistic pietism took when, in the words of Festugière, it became "fatigued with rationalism." It abandoned itself to revelation. The Egyptian touches were in part local color intended to lend mystery and authority.

IV

Among the various topics treated in the Hermetic corpus was astrology. Although in general we shall not consider the fortunes of astrology in this volume, a word about its spread in late antiquity will reveal more clearly the effect of the spiritual currents on scientific endeavor.

Now it must not be thought that astrology was born in Hellenistic or Roman times. For in one form or another it existed as early as the third millennium B.C. in Mesopotamia, and the Greeks themselves were acquainted with it before the time of Plato. We know that Theophrastus composed a work called *On the Signs,* treating Chaldean judicial astrology, but it is now lost. On the testimony of Cicero we know also that the Stoics were well disposed toward the art imported from Chaldea. Among them was Posidonius (about 135 to about 50 B.C.), an astronomer of some note who made a new determination of the circumference of the earth based on a method somewhat different from that attributed to Eratosthenes (see Chap. 7). Posidonius seems to have been addicted to astrology and to have written some five works on it. He made a careful study of the tides and appears to have used the obvious connection of the moon's movement with the tides as evidence for his astrological views. It was Posidonius' account of astrology that St. Augustine studied in his youth.

It is clear that in spite of the long acquaintance of Greek authors with the basic tenets of astrology its rapid spread, which had already been apparent in the Hellenistic period, increased in extent in the first centuries of the Empire. The major objectives of this astrology are to trace the influence of the positions and the movements of celestial bodies on terrestrial activities in general and on those of man in particular, and by a knowledge of these influences, to predict events, including even the character and life of man.

Certain fundamental philosophical ideas lay behind the acceptance of astrology in the Greco-Roman period. Initially Aristotle had given sanction to the influence of the movements of celestial bodies on the terrestrial, or sublunar, world. "This (sublunar) world is tied in some way and in a necessary manner to the local movements of the superior world, in such a way that all the power which resides in our world is governed by these movements" (*Meteorology* I, 4). Yet Aristotle elsewhere accepted chance and undetermined occurrences in our world and he certainly appears to have supported freedom of individual decision. This later came to be the basic difficulty for the Christian astrologers: how to reconcile determinism with the free will of man. The Stoics simply accepted the determinism unqualifiedly.

Among the Stoics and their successors, a related but important philosophical doctrine also acted to support astrology—namely, the doctrine of the unity of the cosmos. From this unity of the ordered universe followed the interdependence of all of its parts, and particularly the dependence of the terrestrial parts on the movement and activity of the celestial. An allied doctrine that was also woven into the general pattern of astrology was that of the identity of the macrocosm (the universe) with the microcosm (man). Man was considered a small cosmos reflecting the larger world. Man, therefore, acted in sympathy with the activity of the universe.

We are particularly interested in the relationship of astrology not only with philosophical doctrines but with the science of the time. From the early days of astrology its liaison with astronomy was close. Astrology naturally had to use astronomy for a determination of the positions and movements of the celestial bodies, for such a determination was the point of departure of the astrologer. Astrology, on the other hand, often provided the means of remuneration for the astrologer-astronomer. The union of astronomer and astrologer in one person was an old one in Mesopotamia, but apparently not in Greece. In all probability such a union was not common until Hellenistic times or later. We have mentioned the astronomer Posidonius, of the first century B.C., who was also interested in astrology. But the classic example was Ptolemy, in the second century of the Christian era. His *Tetrabiblos* was one of the most influential astrological writings of all time. Its influence was felt in late antiquity as well as during the Middle Ages. Part of the work's influence was due to its scientific form. It took up critically the problem of the validity of the astrological art, and the arguments for and against that validity were carefully marshaled.

> Ptolemy next proceeds to set forth the natures and powers of the stars "according to the observations of the ancients and conformable to natural science." Later, when he comes to the prediction of particulars, he still professes "to follow everywhere the law of natural causation," and in a third passage he states that he "will omit all those things which do not have a probable natural cause, which many nevertheless scrutinize curiously and to excess: nor will I pile up divinations by lot-castings or from numbers, which are unscientific, but I will treat of those which have an investigated certainty based on the positions of the stars and the properties of places. . . ." The *Tetrabiblos* has been called "Science's surrender," but was it not more truly divination purified and made scientific? (L. Thorndike, *A History of Magic and Experimental Science*, I, New York, 1923, pp. 112-113.[1])

We shall note only briefly at this point that the rise of astrology did not take place unchallenged. Cicero and other pagan authors had already attacked the doctrine of the Chaldeans before Christian authors opposed it on the ground that it jeopardized the doctrine of free will.

[1] Quoted by permission of the publisher, the Columbia University Press.

V

Another important quasi-scientific product of the spiritual forces of late antiquity was alchemy, the art of transmuting baser metals into gold and silver. The technical details of alchemy we shall examine in a later volume, when considering Islamic alchemy. Now we would like merely to tie up its origins with the movements under consideration in this chapter.

Following one of the ablest students of alchemical history, E. J. Holmyard, we can single out three important currents that came together to produce alchemy sometime in the first three centuries of the Christian era.

(1) The first of these currents is the tradition of the practical working metallurgists, particularly that of the Egyptian metallurgists, for it was apparently in Egypt that the beginnings of alchemy took place. Now we observed in passing in the first chapter that among the technical arts, metallurgy gained an early maturity; methods were developed to reduce metals from their ores, to cast, to form alloys, etc. At some time during the development of metallurgy it became fashionable to color metals artificially so that they would resemble gold or silver. Presumably this imitation of precious metals was already an ancient custom by the time these recipes first began to be written down. We have some of these technological recipes, written on papyrus and dating from the third century. The recipes were not alchemical in nature. They do not pretend to accomplish transmutation. Their objective is simply to produce a product that looks like gold or silver. They are completely free from philosophical or mystical considerations.

(2) If the metallurgical techniques afforded the practical and experimental background to alchemy, it was Aristotle's basic view of the elements and their possible transmutations that provided the theoretical element. As we have already indicated in Chapter 6, Aristotle believed that the elements could be and were continually undergoing transformation, one into another, in nature. Now it was the objective of the later alchemists to see that this transformation took place at the will of man. Processes were sought that would produce the qualitative changes necessary to make changes in elements. More

specifically, the alchemists adopted the Aristotelian view of the nature of metals and minerals found in the *Meteorology*:

> We maintain that there are two exhalations, one vaporous the other smoky, and there correspond two kinds of bodies that originate in the earth, "fossils" and metals. The heat of the dry exhalation is the cause of all "fossils." Such are the kinds of stones that cannot be melted, and realgar, and ochre, and ruddle, and sulphur. . . . The vaporous exhalation is the cause of all metals, those bodies which are either fusible or malleable such as iron, copper, gold. All these originate from the imprisonment of the vaporous exhalation in the earth, and especially in stones. Their dryness compresses it, and it congeals. . . . Hence, they are water in a sense, and in a sense not. Their matter was that which might have become water, but it can no longer do so . . . they all (except gold) are affected by fire, and they possess an admixture of earth; for they still contain the dry exhalation. (Aristotle, *Meteorology*, III, 6, translation of E. N. Webster, Oxford, 1923.[2])

This doctrine of exhalation was to be taken over by the Islamic alchemists and to become the foundation of their theory of metals, the so-called sulphur-mercury theory.

(3) Metallurgy provided the experience and Aristotelian philosophy the chemical theory behind alchemy. But its third element—the essential element, one can say—was the mystical doctrine that formed the cement which bound alchemy into a whole. This element was fed to alchemy by the Hermetic, Gnostic, Neo-Pythagorean and Neo-Platonic literature of the first three centuries of our era. The first of the distinctly alchemical works to possess the mystical and allegorical element dates from about the first century before Christ or the first century after Christ. These were generally attributed to earlier historical figures, such as Democritus, who was renowned for having spent some time in Egypt, and to gods, such as Hermes. These earlier works are referred to by later authors with obvi-

[2] Quoted by permission of the publisher, the Clarendon Press, Oxford, England.

ous reverence, for they seem to give doctrine the sanctity and authority of religious dogma.

One of the first historical alchemists whom we can clearly identify is Zosimus of Panopolis, who lived at Alexandria in the third century. Zosimus occupied a central position in the growth of alchemy, for he unites the earlier tradition with the later Greek alchemy. A number of the later alchemists composed commentaries on the works of Zosimus or were at least inspired by his writings. As we shall see when we examine alchemy in greater detail later, these works contain much of importance for the growth of practical chemical techniques and equipment, even though they possess strong mystical overtones (see Fig. 23). Mystical allegory is particularly evident in the Greek alchemical treatises. One finds Lynn Thorndike's judgment on the allegorical side of these writings repeatedly confirmed as one dips into the great collection of Greek alchemical texts edited by that pioneer student of alchemy P. E. M. Berthelot:

> These early alchemists were also greatly given to mystery and allegory. "Touch not the philosopher's stone with your hand," warns Mary the Jewess, "you are not of our race, you are not of the race of Abraham." In a tract concerning the serpent Ouroboros we read, "A serpent is stretched out guarding the temple. Let his conqueror begin by sacrifice, then skin him, and after having removed his flesh to the very bones, make a stepping stone of it to enter the temple. Mount upon it and you will find the object sought. For the priest, at first a man of copper, has changed his color and nature and become a man of silver; a few days later, if you wish you will find him changed into a man of gold." Or in the preparation of the aforesaid divine water Ostanes tells us to take the eggs of the serpent of oak who dwells in the month of August in the mountains of Olympus, Libya, and the Taurus. Synesius tells us that Democritus was initiated in Egypt at the temple of Memphis by Ostanes, and Zosimus cites the instruction of Ostanes, "Go towards the stream of the Nile; you'll find there a stone; cut it in two, put it in your

hand, and take out its heart, for its soul is in its heart." Zosimus himself often resorts to symbolic jargon to obscure his meaning, as in the description of the vision of a priest who was torn to pieces and mutilated himself. He, too, per-

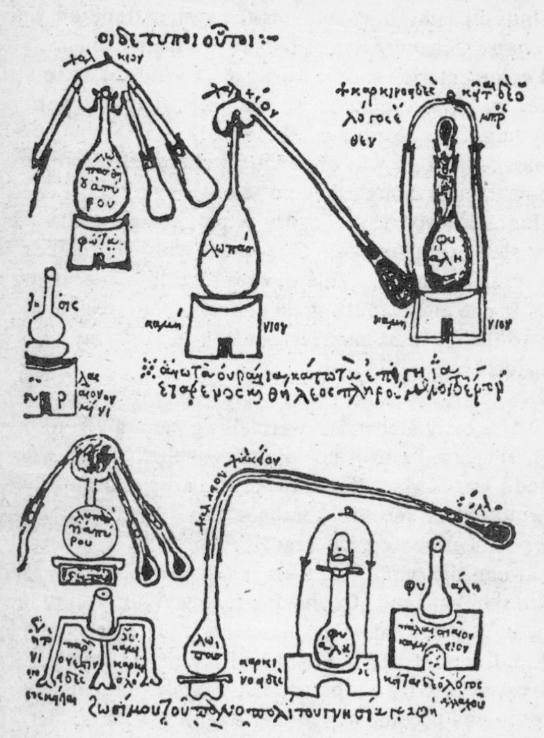

FIG. 23. Greek alchemical equipment appearing in medieval Byzantine manuscripts.

sonifies the metals and talks of a man of gold, a tin man, and so on. A brief example of his style will have to suffice, as these allegories of the alchemists are insufferably tedious reading. "Finally I had the longing to mount the seven steps and see the seven chastisements, and one day, as it chanced, I hit upon the path up. After several attempts I traversed the path, but on my return, I lost my way and, profoundly discouraged, seeing no way out, I fell asleep. In my dream I saw a little man, a barber, clothed in purple robe and royal raiment, standing outside the place of punishment, and he said to me. . . ." [The whole is of course an allegorical description of transmutation.] When Zosimus was not dreaming dreams and seeing visions, he was usually citing ancient authorities.

At the same time these early alchemists cannot be denied a certain scientific character, or at least a connection with natural science. Behind alchemy existed a constant experimental progress. "Alchemy," say Berthelot, "rested upon a certain mass of practical facts that were known in antiquity and that had to do with the preparation of metals, their alloys, and that of artificial precious stones; it had there an experimental side which did not cease to progress during the entire medieval period until positive modern chemistry emerged from it." (Thorndike, *A History of Magic and Experimental Science*, I, pp. 197-198.[3])

VI

In discussing the rise of alchemy we made reference to Neo-Pythagoreans and Neo-Platonists. The rise and spread of their interrelated philosophical movements is further testimony to the strength of the spiritual forces loose in the Empire. It is most difficult to order the miscellaneous information that remains of the Neo-Pythagorean movement. We see traces of this movement from the first century B.C., and it is not impossible that there were tenuous ties of the Neo-Pythagorean movement with the remnants of the older Pythagorean movements of the fifth century B.C. The greatest difficulty of all is experienced in trying to separate doctrines of the Neo-Pythagoreans from those of the Pythagoreans of the early period, because of the tendency to ascribe everything to the master himself as a kind of a prophet. In fact, in many ways the Neo-Pythagorean movement appears to us more as a church than as a philosophical school. Discussion among the members was terminated by the almost ritualistic expression *ipse dixit* (or, in

[3] Quoted by permission of the publisher, the Columbia University Press.

Greek, *autos epha*) — i.e., "He (Pythagoras) has declared. . . ." There existed a Pythagorean catechism. Pythagoras' doings, like the later *acta* of the Christian saints, were recorded, and included among them were miracles. Sacred books were attributed to him. Hymns were composed. In the course of the development of the movement Homer and Plato were annexed to it. The works of both were the object of exegesis to reveal their spiritual messages.

The importance of mathematics and particularly of number theory to the Neo-Pythagorean movement apparently caught the imagination of a number of mathematicians who associated themselves with the movement. We have already mentioned the rather mediocre mathematician of the first century, Nicomachus, who is said to have been a Pythagorean and to have composed, in addition to his manuals on arithmetic and harmony, a work on the "theology" or mystic properties of numbers.

This devotion to mathematics is also apparent among the Neo-Platonists, who undoubtedly had ties of varying firmness with the Neo-Pythagorean movement. We cannot possibly detail the philosophic content of this movement, so irritating at times and yet so important for an understanding of the intellectual history of the Middle Ages; but its relationship with contemporary and medieval science should not escape us. Among the early Neo-Platonists Plotinus should be mentioned, who flourished from about 203 to 270. Historians of philosophy concern themselves with Plotinus' development of the characteristic Neo-Platonic doctrine of the emanation and hierarchy of beings and powers, this emanation proceeding from the First Being through spiritual or noncorporal beings to matter. The influence of the spiritual forces of the times is reflected in the importance given to incorporeal beings, such as Intelligence (*Nous*) and Soul (*Psyche*) in this hierarchy, and in the assumption of matter as the source of evil. It should be remarked that the mystic excesses founded in the works of some of the later Neo-Platonists are missing in the works of Plotinus. He was at times quite critical of the occult science of his day. Thus he asserted that diseases are due not to demons but to natural causes. It is true that he accepted the effects of magic on the physical nature of man. "But the rational soul may free itself from all influences of magic. . . . The life of reason is alone free from magic" (paraphrase of Thorndike).

Plotinus' amendment to astrological doctrine is worth noting. The movements and positions of stars are generally not the causes of events but are merely signs or indications of future events, although in some cases they are both the causes and the signs. They operate as signs because the First Principle has established a perfect harmony among the manifold things that make up the universe. On the other hand, "Plotinus made a distinction between the extent of the control exercised by the stars over inanimate, animate, and rational beings. The stars signify all things in the sensible world but the soul is free unless it slips and is stained by the body and so comes under their control. . . . Plotinus arrives at practically what was to be the usual Christian position in the middle ages regarding the influence of the stars, maintaining the freedom of the human will and yet allowing a large field to astrological prediction." (Thorndike, *A History of Magic and Experimental Science*, I, p. 306.[4])

Among the successors of Plotinus, Iamblichus, who died about 330, is of some importance because of his interest in mathematics and his close ties with the Neo-Pythagoreans. He composed some nine books on the Pythagoreans, including a long life of Pythagoras, a commentary on the *Introduction to Arithmetic* of Nicomachus, and books on arithmetical science in physics, in ethics, and in theology. He was well acquainted with Pythagorean number mysticism. He is also credited with a short chemical tract.

Also important for the history of science were the late Neo-Platonists Proclus, Simplicius, and Philoponus. The last two, who flourished in the first half of the sixth century, played an important role in the growth of the criticism of Aristotelian ideas, particularly in mechanics. And their influence among the Islamic philosophers and the later medieval authors was significant. Hence we have reserved a separate chapter for them. Proclus (410-485) was perhaps the ablest of the Neo-Platonic mathematicians. He was trained at both Alexandria

[4] Quoted by permission of the publisher, the Columbia University Press.

and Athens. In philosophy his renown rested upon his numerous commentaries on the dialogues of Plato. Apparently Proclus taught mathematics at the Neo-Platonic school at Athens. One of the products of this teaching, no doubt, was his *Commentary* on Book I of Euclid's *Elements*, a work rich in historical references. Proclus also composed a work called *Outline of Astronomical Hypotheses*, which was an introduction to Hipparchus and Ptolemy.

That Neo-Platonism was responsive to the times is demonstrated by the fact that this was the philosophy which was to have the most influence on the early Christian writers as they composed their apologetic and dogmatic works. To complete the chronicle of the effects of the spiritual forces on intellectual and scientific activity we must examine that Christian literature.

Chapter 10

Science and Patristic Literature

I

WE HAVE singled out Christianity as the most important and obvious result of the triumph of the spiritual forces that beset the Empire in the first three centuries after Christ. We have also noted that in its growth and triumph the Church siphoned off men who might well have pursued natural philosophy or science. Then we suggested, without documentation, that the fathers interpreted science according to their predilections for moral philosophy and the predispositions that naturally came from their sacred literature.

In this chapter we must provide some of the necessary documentation by examining more closely the attitudes prevalent among the Church fathers toward Greek philosophy and science and then by following this examination with a brief account of some of the common scientific and philosophic opinions held by the fathers.

But before we undertake a more detailed analysis of the attitudes and opinions of the fathers, we should note that the Christian thought of late antiquity owed much to the fundamental attitudes and opinions of the Jews who flourished in Alexandria from the time of the translation of the Pentateuch or Books of Moses into Greek, about 260 B.C., until the first century after Christ. Professor Harry Wolfson, of Harvard University, has shown that the attempt to use Greek philosophy by these Jews of Alexandria, and particularly by their outstanding philosopher Philo Judaeus, whose activity extended until about A.D. 40, resulted in the elaboration of some basic philosophical and religious views that were stamped on the whole succeeding religious philosophy of the late antique Greek Christians, of the Syriac Christians, of the Islamic and Jewish medieval philosophers, and finally of Latin medieval schoolmen.

The dominant position of revealed truth as opposed to

rational doctrine, with the relegation of philosophy to the position of a handmaiden to theology, the essential unity of correctly interpreted revealed truth with correctly reasoned truth, the necessity of allegorical as well as literal interpretation, to say nothing of fundamental religious views that do not concern us here, such as the existence and unity of God—all these basic attitudes and doctrines appear in scattered fashion in the writings of Philo and were made the basis of the medieval religious philosophy. And as we examine the attitudes and opinions found in Patristic literature, we shall more than once note their prior exposition by Philo.

If the fathers were to bend philosophy to the use of Christianity, they first had to show wherein the philosophers, who were for the most part, they claimed, ignorant of the Scriptures or had distorted them, had fallen into error. Hence, even though philosophy was to be used as a handmaiden by the fathers, they nevertheless often criticized the philosophers with distinct hostility. One of the favorite charges against the Greek philosopher by the fathers was that he had gotten the truth, on occasion from the prophets, but for pride and vainglory had distorted it. The Latin father Tertullian, who flourished about A.D. 200, in his *Apologeticus* (Chap. 47), a defense of Christianity, expressed this view. He asks what poet or sophist has not drunk at the fountain of the prophets? And he asserts that "the philosophers watered their arid minds" from the Scriptures. But the philosophers sought only glory and eloquence, and "if they fell upon anything in the collection of sacred Scriptures which displeased them, in their own peculiar style of research, they perverted it to serve their purposes." The hostility of Tertullian was also demonstrated in the work *Ad nationes* (II, 2), where he endeavors to show that the natural philosophers (*physici*) were in error, as well as the poets and others. Now all things with the philosophers are uncertain, because of their variation. Philosophy has not recognized that a man who has the fear of God and thus knowledge of the truth of God has "full and perfect wisdom." And even the truth which the natural philosophers had discovered "degenerated into uncertainty, and there arose from one or two drops of truth a perfect flood of argumentation (*argumentationum inundatio*)."

The argumentation, cantankerous disputes, and disagreement of the philosophers on fundamental matters were often alluded to by the fathers and contrasted with the supposed unity of Christian belief. Thus St. Basil (329-379), in noting the dispute that arose among the philosophers over the nature of celestial bodies, remarks, "If we attempt to treat here of these questions we shall fall into the same pitfalls as they (the philosophers). Let us leave them to ruin themselves and to refute each other" (*Hexameron*, I, 11). In another place the same author compares the simplicity of spiritual discourse to philosophical discussion. "As the beauty of the chaste woman surpasses that of the courtesan, so our discourse is superior to those of men who are strangers (to Christianity)" (*Hexameron*, III, 8). St. Augustine (354-430) expresses the same idea in his *City of God* (XVIII, 41): "As for our canonical authors, God forbid that they should differ. . . . (But) let one look amongst all the multitude of the philosophers' writings, and if he find two that tell both one tale in all respects, it may be registered for a rarity."

But hard as Augustine is on the philosophers he does admit that there is certitude in the work of the astronomers, who seek out things by the nature and light of the spirit that God has given them. They predict several years in advance solar and lunar eclipses, giving the day, the hour, and the magnitude of the eclipses. And, he goes on to say, that which they have predicted takes place just as predicted (*Confessions*, V, 3).

Some of the hostility—one might better say indifference—to philosophy stems from the fact that the philosophers, even where they are not wrong, are missing the whole point of existence, the discovery of God. So St. Basil castigates the astronomers most of all for being so wrapped up in the vanities of their research that they fail to comprehend the important truth. They fix the locations of the stars in the northern and southern skies; they observe with great care the retrogradations, the stations, the declinations, and other movements of the heavenly bodies; they determine the periods of the planets; and so on.

But they miss one thing, "the discovery of God, the Creator of the universe, the Just Judge" (*Hexameron,* I, 4). In like fashion Augustine in his *Enchiridion* insists that the opinions of the *physici* are of little moment for the Christian:

> When . . . the question is asked what we are to believe in regard to religion, it is not necessary to probe into the nature of things, as was done by those whom the Greeks call *physici*; nor need we be in alarm lest the Christian should be ignorant of the force and number of the elements—the motion, and order, and eclipses of the heavenly bodies; the form of the heavens; the species and the natures of animals, plants, stones, fountains, rivers, mountains; about chronology and distances; the signs of coming storms; and a thousand other things which those philosophers have either found out, or think they have found out. For even these men themselves, endowed though they are with so much genius, burning with zeal, abounding in leisure, tracking some things by the aid of human conjecture, searching into others with the aid of history and experience, have not found out all things; and even their boasted discoveries are oftener mere guesses than certain knowledge. It is enough for the Christian to believe that the only cause of all created things, whether heavenly or earthly, whether visible or invisible, is the goodness of the Creator, the one true God; and that nothing exists but Himself that does not derive its existence from Him; and that He is the Trinity—to wit, the Father, and the Son begotten of the Father, and the Holy Spirit proceeding from the same Father, but one and the same Spirit of Father and Son. . . . (*Enchiridion,* Chap. IX, translation of J. F. Shaw.)

In another place (*De Genesi ad literam,* Chap. 10) Augustine, in replying to some detailed cosmological question, claims that the answers take detailed investigation for which he has not the time. And this kind of activity is not useful and necessary for the Church.

If the fathers were often hostile or indifferent to the activity of the philosophers, it was admitted, as the initial passage quoted from Tertullian illustrates, that they occasionally did get hold of the truth. But when they did, many of the fathers held, that truth came ultimately from God. This truth was possessed first by the prophets and is enshrined in the Scriptures. The truth revealed by God was, when properly interpreted, true wisdom of a kind superior to results of human reason. This attitude had prevailed among the Jews, as was reflected in the works of Philo and was reiterated a century or more later by the early Christian father Justin (*ca.* 150). Justin tells us that in ancient times there were "happy men, just men, cherished by God, who spoke by the Holy Spirit." These prophets made oracles which have come to pass; they saw and announced the truth to man. Those of their writings that still exist can be examined with profit by the philosopher who seeks knowledge on the beginnings and end of things, or on any other matter. "But it is not by demonstration they have advanced their statement, for they are witnesses of the truth beyond any demonstration" (*Dialogue with Trypho,* Chap. 7).

This concept of truth by revelation, one of the paramount attitudes present in the medieval writings, we have now seen to have existed in the Hermetic writings, in the semiphilosophical works of the Jews of Alexandria, and then in the works of the Christian fathers. We have insisted that it constitutes a fundamental alteration from the spirit of Greek rationalism. And the best science of the medieval period is that which ignores or at least minimizes revealed truth in favor of agreement with empirical evidence or rational cogency.

But if the fathers thought of revealed truth as the more certain knowledge, the early Christian father, and Philo before him, often made the important supplementary assumption that there is really only one truth, and properly conducted philosophy or human reason will also arrive at that truth. This attitude has been called the double-faith doctrine, when applied to truth of God; faith can be achieved by both revelation and reason. The reader familiar with medieval scholastic writings will recognize this as a fundamental view of St. Thomas Aquinas.

Listen to Clement of Alexandria (*fl. ca.* 217) in his *Miscellanies* as he talks about the gnostic, the seeker after truth.

> For him knowledge (*gnosis*) is the principal thing. Consequently, therefore, he applies to the subjects that are a training for knowledge, taking from each branch of study its contribution to the truth. Prosecuting, then, the proportion of harmonies in music; and in arithmetic noting the increasing and decreasing of numbers, and their relations to one another, and how the most of things fall under some proportion of numbers; studying geometry, which is abstract essence, he perceives a continuous distance, and an immutable essence which is different from these bodies. . . . And by astronomy, again, raised from the earth in his mind, he is elevated along with heaven, and will revolve with its revolution. . . . Further the gnostic will avail himself of dialectics, fixing on the distinction of genera into species, and will master the distinction of existences till he comes to what are primary and simple. But the multitude are frightened at the Hellenic philosophy, as children are at masks, being afraid lest it lead them astray. But if the faith (for I can not call it knowledge) which they possess be such as to be dissolved by plausible speech, let it be by all means dissolved, and let them confess they will not retain the truth. (*Miscellanies*, Book VI, Chap. 10, translation of William Wilson, New York, 1903.)

Clement goes on to say that the righteous man will not dread cunning words if he is able to distinguish them or answer them correctly. Logic, then, becomes a bulwark so "that truth cannot be trampled under foot by the sophists." "The studies of philosophy, therefore, are aids in treating of the truth." Clearly, Clement has stressed the basic idea that was to persist from the time of Philo through a good part of the Middle Ages, that philosophy is the handmaiden to religious truth—to theology, if you will. The handmaiden idea goes back at least to Philo, who notes that just as the encyclical studies (grammar and other liberal arts) are handmaidens to philosophy, as the Stoics and others wished, so philosophy is the handmaiden of wisdom — i.e., religious truth. Philo actually uses the term "handmaiden" (*therapainis*).

Origen (*ca.* 185-254), Clement's junior contemporary at

Alexandria, expresses this ancillary concept in much the same fashion when he urges a friend to direct the whole force of his intellect to Christianity as his end:

> And I would wish that you should take with you on the one hand those parts of the philosophy of the Greeks which are fit, as it were, to serve general or preparatory studies for Christianity, and on the other hand so much of geometry and astronomy as may be helpful for the interpretation of the Holy Scriptures. The children of the philosophies speak of geometry and music and grammar and rhetoric and astronomy as being ancillary to philosophy; and in the same way we might speak of philosophy itself as being ancillary to Christianity. (*Epistle to Gregory*, Chap. 1.)

Note that, of the disciplines mentioned by both Clement and Origen, the sciences of the quadrivium—arithmetic, geometry, music, and astronomy—stand out as the natural sciences. These are the sciences first emphasized and grouped together by the Pythagoreans and then pursued by the Platonists.

Although we have noticed in Augustine's writings some impatience with secular philosophical studies as not leading to God, there is a clear case of his utilizing the ancillary concept. In one place in his *Confessions* (V, 3) he refutes certain Manichean (the Manicheans were a quasi-Christian sect in the Near East) ideas about astronomy by an appeal to the astronomy he had learned following the rules of mathematics and his own eyes. The point of Augustine's argument is that, even though Christianity rather than science is wisdom, if we can show by an appeal to astronomy that the Manicheans are wrong in their science, they are no doubt wrong about their religion as well.

The persistence of the subsidiary position of science and philosophy to theology has been pointed out as having an important effect on philosophical activity in the Middle Ages. But further clarifying ideas might well be observed. The first is that it was always difficult to say where the study of philosophy as ancillary was to stop. A man with a philosophical bent—an Abelard, say, or an Ockham—might fundamentally agree that philosophy is ancillary, but as a rather independent philosopher he pursues philosophy across any arbitrary boundaries between philosophy and theology. And, what is more important in the history of science, the decision to utilize Greek philosophy as ancillary brought about the further study of the sciences that were in a certain sense ancillary to philosophy. Let us take, for example, the decision made at the University of Paris that logical works of Aristotle could be included in Christian studies. Following this decision there was increased pressure for the study of Aristotle's works in natural science, and soon they were widely studied. A final point worth noting is that, in spite of the continued affirmation of the subsidiary nature of secular scientific studies by the natural philosophers, there were always some who pursued the Greek scientific tradition without consideration for an ulterior religious purpose and outside the confining limits of the superiority of revelation. We hope to show later the virility of the Hellenistic tradition in the thirteenth century in mathematics, mechanics, and other subjects, which was in a sense independent of the religious philosophy of the day.

II

We have spoken of some of the general attitudes of the Christian fathers toward Greek philosophy and science. Now let us examine some of the particular philosophical and scientific opinions that appear in their apologetic and exegetic writings.

The fathers were led to a discussion of natural philosophy by the apparent differences and similarities between the account of creation given in Genesis and the opinions of the philosophers. Commentaries on Genesis go back at least to Origen (*ca.* 185-254), and in Jewish writings, of course, much further—at least to the important *De Opificio mundi* (*On the Creation of the World*) of Philo the Jew in the first century or earlier. Of Origen's commentary we possess some interesting fragments, one of which recounts the rather recondite theory of the astronomers on the precession of the equinoxes. This phenomenon, we may recall, was discovered by Hipparchus and discussed by Ptolemy, who lived in Alexandria scarcely more than a generation earlier than Origen.

As commentaries on Genesis became more popular, the so-called *Hexameron* form of commentary arose. As the name indicates, the *Hexameron* discussed the creation activity of the six days. Perhaps the most important of these *Hexameron* writings were the *Homilies on the Hexameron* of St. Basil (329-379), who had been trained in the schools of Alexandria and Athens. Basil's *Hexameron* was defended and elaborated by his younger brother Gregory of Nyssa (*ca.* 330-400), and it was paraphrased by one of the Latin fathers, St. Ambrose (340-397). It was also translated into Latin in the fifth century and was widely read during the early Middle Ages, as its further epitome into Anglo-Saxon illustrates. And one of the most important of the English churchmen of the early Middle Ages, Bede, whose activity we shall discuss in a later chapter, utilized it in part in his *Hexameron.* St. Augustine (354-430) also composed some commentaries on Genesis, the most important for our purpose being his *De Genesi ad literam.* But in general we shall make greater use of his more widely known works, the *Confessions* and the *City of God.*

We should point out immediately that the scientific as distinct from the more general philosophical ideas expressed by these authors were of a very elementary nature. None of their works, with the exception of the sixth-century *De Opificio mundi* of the former Neo-Platonist John Philoponus, shows the expert's eye. As Duhem points out, "the science which they suppose their listeners and readers to know and with which they themselves seem to be content is composed of a number of simple and general propositions." These writings were addressed, then, to the general run of Christians, who could not be expected to be familiar with the more advanced doctrines of the scientists.

One of the fundamental views taken over by the fathers from Greek philosophy, but radically modified in the Christian writings, was that there exists in nature a generally fixed order expressible in terms of laws, for the most part immutable. This fixed order with its laws may, however, be set aside in miraculous fashion at God's pleasure. The introduction of the idea of the miracle, of God's complete freedom of action with respect to the natural order of things, was a distinct modification of the Greek concept of natural order. In their general attitude, Philo and the Jews of Alexandria were perhaps the first to give philosophic expression to the concept of the natural order modifiable at the will of God. Philo had noted that "the nature of the heavenly bodies and the movements of the stars . . . the vast number of other operations in nature . . . are invariably carried out under ordinances and laws laid down in the universe as unalterable" (*De Opificio mundi,* XIX, 61; cf. H. A. Wolfson, *Philo,* I, Cambridge, Mass., 1947, p. 328). And Philo goes on to describe three such fundamental laws: the law of the existence of opposites, the law of the harmony of opposites, and the law of the perpetuity of species. Of the last law he says: "God willed that nature should run a course that brings it back to its starting point, endowing the species with immortality, and making them sharers of eternal existence," by the very cycle of life from plant to fruit to seed to plant again. But, although these laws are generally immutable, God as their author can suspend them, reverse them, or operate in any fashion he desires outside of them. Wolfson summarizes the pertinent remarks of Philo on this subject:

> While admitting that earth, rain, air, husbandry, medicine, and marriage in the ordinary course of nature are productive of certain effects, yet he (Philo) maintains that "all these things, through the power of God, admit of change and transition, so as often to produce effects quite the reverse of the ordinary." (Wolfson, *Philo,* I, Cambridge, Mass., 1947, p. 349.)

The Christian authors speak in precisely the same way and to a man. Thus Augustine answers the protestation that miraculous changes in things are contrary to their nature by asserting that God is the author of all nature and so the miracle is simply "by the Will of Almighty God." However, Augustine admits that if the alleged miracle is not recorded in the Scriptures we have a right to be skeptical. Some at least unnatural or curious things he himself has proved by trying, among them being "the burning of lime in water and cooling in oil; the loadstone's attraction of iron, but not straw," and so on.

Other authors express suspicion of some of the miracles recorded but none doubted the possibility of miracle through God's omnipotence. One effect of such a doctrine was to encourage even very good practicing scientists, such as Albertus Magnus, to be somewhat more credulous, although credulity is by no means an attribute of any particular religion or society, as Professor Thorndike's research into the history of magic and experimental science has well shown.

Although the fathers believed generally in the laws of nature as modified by possible miraculous intervention, for the most part they shied away from interpreting the actions of laws in a completely deterministic fashion; for not only was God free to change the natural order when he wished, but man was free to exercise his will. Thus most of the fathers expressed determined opposition to the determinism implied in astrology.

Among the earliest opponents of astrology was the already mentioned Tertullian. Astrology he believed to be an invention of the fallen angels, as were also botany and metallurgy. And Basil noted that the acceptance of astrology would obviate the great hopes of the Christians and the whole system of justice with attendant punishments and rewards.

Most interesting of the criticisms of astrology was that developed by St. Augustine in the fifth book of his *City of God*. The casting of horoscopes comes under his particular criticism. If the principle of like causes producing like effects is correct, and the movements and positions of the stars at the time of birth determine or even signify the future activities of man, how can we account for the different lives and characters of twins, even though they would have the same horoscopes? He describes the varied activity of twins he knows; one is a staff officer (V, 6), married and with a number of children, continually at war; the other, his sister, a holy virgin, who never leaves her country. Similarly, think of the diverse futures of seeds sown under the same astrological conditions (V, 7).

If the Christian writers were vigorous opponents of the restriction that astrology would place on the human will, they nevertheless often accepted the limited influence of the heavenly bodies on transformations of sublunar matter. And

Augustine gives sanction to this view when he admits that "it is no such absurdity to say that there are some planetary influences that have effect only upon diversity of forms in bodies, as we see the alteration of the year (the seasons) by the sun's access and departure, and divers things to increase and decrease, just as the moon does (crabs, for example, and all shell fish: besides the wonderful course (tides) of the sea) yet it is absurd to say that the mind of man is subject to any of these powers of the stars" (*City of God,* V, 6). Augustine does admit that the astrologers often foresee many wonderful and true things, but they do so with the help of evil spirits whose intention is to deceive man into accepting "this false and dangerous opinion of fate in the stars" (*Ibid.*, V, 7).

In addition to introducing the miraculous action of an almighty God as a serious modification in the concept of the natural order, and to freeing the activity of man from the influences of celestial bodies, the fathers, in line with the account in Genesis, opposed the general philosophical view that held matter to be eternal (see Chap. 3 above). The opening sentence of Genesis indicates that "In the beginning God created the heaven and the earth." Some of them admit that he made the earth out of pre-existent matter—as Augustine calls it, "a certain unshapedness"—but even that must have been created by God. The fathers tend to believe that the ambiguous account in Plato's *Timaeus* supports the creation doctrine. Some appeal, certainly with no justification, to the Aristotelian concept of the Prime Mover, who appears to be coeternal with matter.

Holding to the creation doctrine, the fathers also affirmed that time had a beginning when the world was created. The fundamental connection of time with movement, recognized by Plato, Aristotle, and the Stoics, was usually acknowledged by the fathers when they had anything to say on the subject of the nature of time. Even before the fathers, Philo had told us that "time was conceived when bodies were in motion," and that the world of our senses, when set in motion, has caused the nature of time to shine forth and to become conspicuous." St. Augustine in his *City of God* stresses the necessary relationship between time and movement:

> For if eternity and time be rightly distinguished, time never to be extant without movement, and eternity to admit no change, who would not see that time could not have being before some movable thing were created; whose motion and successive alteration (necessarily following one part after another) the time might run by? (*City of God*, Book XI, Chap. 6.)

However, Augustine submitted the problem of time to a much more profound analysis in his *Confessions* (Book XI). Although time units are related to the movements of heavenly bodies, such movements do not constitute time. "Time is not the motion of a body" (XI, 24). Time is the duration in which these movements take place and by which they are measured. Accentuating the difficulty as to what is time is the realization that past time no longer exists, future time does not yet exist, and present time as something instantaneous has no duration. Yet we do affirm that past, present, and future exist. How do they exist? They exist in the mind. "The present time of past things is our memory; the present time of present things is our sight; the present time of future things is our expectation" (XI, 20). As to the measurement of time, Augustine says (XI, 27), "It is in thee, O my mind, that I measure my times. . . . In thee, I say, it is that I measure times. The impression, which things passing by cause in thee, and which remains when the things are gone, is that by which I measure time, rather than the things which passed by to produce the impression." It is also in relation to some impression that an anticipated duration in the future is measured.

Augustine has, then, given a rather nice psychological analysis of time which supplements the earlier treatments of time relating it to movement.

Once having established in the beginning of their treatments of Genesis that matter and time were created and thus had beginning, the fathers usually fell back on earlier Greek ideas as to the nature of the elements, particularly the ideas of Plato and Aristotle. The doctrine of four elements, each with its pair of qualities, which Aristotle developed out of earlier ideas, was quite widely accepted by the fathers. Thus St. Basil in his *Hexameron* follows Aristotle rather closely in his acceptance of the four elements, each with its natural place; similarly, his analysis of natural movement is strictly Aristotelian. Accepting the qualitative nature of the elements as outlined by Aristotle, Basil recounts (IV, 5) a contemporary doctrine of how compounds are formed, stating that it is only compounds which we see and not the elements in pure state. The unions or mixtures of elements take place by the sharing of a quality common to each of them. Thus water can combine with earth because they are both cold, and water can combine with air because they are both moist, etc. The sharing of a common quality, then, was believed to be the mechanism of union. This doctrine, as Duhem has shown, had some popularity in the Middle Ages.

When it came to the nature of the substance of which celestial bodies were formed, the fathers followed no single position. Basil refuses to decide whether they are composed of some mixture of the four elements or of the fifth element. But Basil's brother, Gregory of Nyssa, takes the Platonic position that the heavens are composed of fire. When the fire mounts to the upper regions, it necessarily turns in a circle upon reaching the upper regions. Similarly, Ambrose prefers a position holding the heavens to be composed of fire and water rather than of the Aristotelian fifth essence.

Accepting the concept of four elements, the fathers came up against the problem of prime matter, the material substratum the independent actual existence of which the Aristotelians had denied in favor of a potential existence only. Under Neo-Platonic influences, certain of the fathers transformed this doctrine into the notion that prime matter can exist in actuality.

> St. Basil assuredly had an idea of prime matter completely different from that which Aristotle conceived and quite analogous to that which we shall see St. Augustine borrowed from the Neo-Platonists. He takes the word *hylê* (matter) in a sense very close to that which we give the word "matter" to-day. He sees in it a body, but a body incompletely and badly defined. As to that matter deprived

of all form and of every quality of which the philosophers speak, he would seek it only as pure denial, entirely inconceivable. (P. Duhem, *Système du Monde,* II, Paris, 1913, p. 429.)

St. Augustine seems to have a doctrine not unlike that of Basil, in which prime matter emerges as an incompletely defined matter. He notes in his *Confessions* (XII, 3) that it cannot be denied completely, that it must have some existence, perhaps as "unshapedness." This concept of actual prime matter was present among the Neo-Platonists. In an entirely different fashion, another concept of actually existing first matter existed among the atomists, for whom it was the atoms stripped of all secondary qualities caused by their configuration. Possibly both traditions of the actual existence of prime matter were important in the formation of the modern concept of mass.

Of the cosmological ideas forced upon the fathers by the account of Genesis none was more contrary to the teaching of Greek philosophy than the idea that there existed waters above the firmament. Origen interprets this doctrine of supracelestial waters allegorically, but many of the fathers take it seriously. The most sensible of the accounts of those who accept the doctrine is that of John Philoponus, the sixth-century Christian and former Neo-Platonist. According to Philoponus, Moses was not concerned with the technical considerations of astronomy and cosmology. His principal aim was to lead men to an understanding of God, and Genesis should be examined in this light. Philoponus, of course, had a far better understanding of Greek science than any of the other fathers we have had occasion to cite, and hence his account is richer in allusion to the prior opinions of the philosophers. He believed that Moses' account of the nature of the firmament was more satisfactory than that of Plato and Aristotle, which Philoponus had cause to review briefly. He thinks that Moses, because of the transparency and fluidity of water and air, believed that a solid or rigid substance had been formed still translucent; this solid substance was the firmament. "Moses, then, has suggested the idea that the Heaven, because of its transparency, was formed in major part from the air and water; this idea is more physical and accords better with appearances than the hypotheses of Plato and Aristotle" (*De Opificio mundi,* III, 5). Of course, he goes on to say, the heavenly bodies, obviously not transparent, are formed of a different igneous substance than the surrounding transparent firmament. He dismisses the Aristotelian concept of a fifth element.

We have talked mostly about the cosmological ideas of the fathers, and necessarily so, because the fathers centered their interest on that part of philosophy and science which would contrast with the ideas expressed in Genesis. But one field of scientific endeavor in which the fathers occasionally expressed detailed scientific opinions was natural history. Direct use was made of the habits of animals to reveal Christian dogma or morals. The result was a Christianized natural history. There is little evidence among the various authors who write such natural history that they themselves did any independent investigation. Ordinarily they give a description of the habits of the animal taken either from folklore or, often in truncated or incorrect form, from such scientific works as the zoological writings of Aristotle; this is followed by a moralization based on Christian principles. These moralizations go much beyond the rather modest moralizing we find in the *Natural Questions* of the Roman author Seneca. But it is supposed, on the basis of those questions of Seneca and other evidence, that such a bastardizing of natural science was by no means exclusively Christian. As a matter of fact, Aristotle's works themselves are by no means completely free of such moralizing.

Earliest of the works that illustrate the Christianizing of natural history is the so-called *Physiologus,* which treats of forty-eight actually existing or mythical animals, plants, and stones. It stresses their sometimes real but often fabulous properties in order to make some moral regarding Christ, the Devil, or the Church. It draws its name from the continued repetition of the phrase *physiologus dicit* or its equivalent in Greek. Thus the description of the properties or habits is introduced by the phrase "The Physiologue says." Who the original physiologue—which simply means "student of nature" —was, we do not know. We know that the work is cited unambiguously by fourth-century fathers, and it may go back to the second century. It has been conjectured that it was written at Alexandria in Greek. But there are versions in numerous languages, including Greek, Latin, Armenian, Ethiopic, Syriac, and the later European languages.

One author has attempted to show that, like many of the romantic-mystical writings in natural history and allied topics of the first centuries after Christ, the *Physiologus* goes back to one Bolos of Mendes, who flourished about 200 B.C. Bolos composed a work on powers, animals, and stones; it was entitled *Natural Potents* and had subtitles *Sympathies* and *Antipathies*. It is not impossible that Pliny's *Natural History*, of the first century, had some influence on the *Physiologus* and similar products of moralized natural history. The account of the viper in *Physiologus* illustrates the type of material and form of the work:

> John said to the Pharisees: A generation of vipers (Matt. 3, 7; Luke, 3, 7). Physiologue says concerning the viper that the male has the appearance of a man, while the female has the figure of a woman up to the umbilicum, but the figure of a crocodile from the umbilicum up to the tail. Moreover, the female does not have any hidden place, i.e. any hollow member for bearing (children), but has a pointed hole. Further, if the male cohabits with the female, he ejects semen into the mouth of the female, and if she drinks his semen she cuts off the necessary or virile parts of the male, and the male dies. When, moreover, his children are born in the uterus of their mother, who does not have any bearing pouch, the children rip open the side of their mother and come out killing their mother.
>
> Hence our Saviour assimilated the Pharisees to the viper, for this generation killed its father and mother, and thus here the people who are without God kill their father Jesus Christ and their earthly mother Jerusalem. And how

will they flee from the wrath to come? Moreover, our Father Jesus Christ and Mother Church will live unto eternity, while they living in sin are mortal. (*Physiologus,* Chap. XIII, translated from Latin version y, edited by F. J. Carmody.)

Of a somewhat higher scientific character and interest was the natural history that appeared in some of the *Hexameron* treatises, particularly those of Basil and Ambrose. These works have some scientific importance, because they rely in part on the natural history of Aristotle, and they constitute the only form of some of Aristotle's zoological ideas in the early Latin West before the great period of translations of the twelfth and thirteenth centuries. Basil, for example, in the last of his homilies is either directly or indirectly dependent on Aristotle, not only for some of his statements on the classification of animals but also in many cases for the specific factual knowledge about their habits. Thus, in talking about the necessity of air for insects (VIII, 7), he makes the same observation that Aristotle made, that if the insects are covered with oil they will not survive. Basil goes one step beyond Aristotle by saying that they do not have lungs, but rather breathe through the pores of their whole body, and these pores will be covered when we dip them into the oil. The distinction of viviparous from oviparous marine animals obviously comes from Aristotle, as do his remarks on the similarities and difference between these basically different animals. Basil makes an interesting observation that it is a law of nature (*nomos physeōs,* or *naturae lex*) "which according to the needs of each kind (of fish) has allotted to them their dwelling places with equality and justice" (*Homilies on the Hexameron,* VII, 3).

Although the *Hexameron* of the early father does not ordinarily have the fable or folk quality of the *Physiologus,* it does utilize the habits of animals for the illustration of morals. Hence Basil after keeping fairly close to the natural history of marine animals for three chapters, proceeds to moralize for a number of chapters. Here is a sample of that moralizing:

> A fish does not resist God's law, and we can not endure His precepts of Salvation! Do not despise fish because they are dumb and quite unreasoning; rather fear lest, in your resistance to the disposition of the Creator, you have even less reason than they. Listen to the fish, who by their actions all but speak to say: it is for the perpetuation of our race we undertake this long voyage. (Basil, *Homilies on the Hexameron*, VII, 4.)

As these samples of Christian cosmological opinion and natural history amply illustrate, the objective of the fathers of the church was fundamentally different from that of the Greek philosopher or scientist who made natural philosophy his profession. But even though he carried on little if any independent scientific research (Philoponus being a notable exception), the early Christian writer often illustrated a persistent interest in and wonder of nature that made him seek out scientific opinion of the past and transmit it, however he may have perverted and obscured it. And surely as Christian Europe matured after many centuries of political and economic settling down, these Christian writings whetted the appetite of the student of the eleventh and twelfth centuries, making him seek out the great body of Greek and Islamic learning possessed by Islam.

Chapter 11

Latin Science in Late Antiquity

I

UP TO THIS point we have studied the scientific activity of the antique world very much as a whole. Now, in conformity with the events of history, we must begin to consider Western Latin science separately from the main stream of Greek scientific development in the Eastern part of the Roman Empire.

The chief factor dictating an increasingly separate and independent development in the West was the relative success of barbarian incursions in the West. From the early days of the Empire (and even before) the pressure of the barbarians on the borders of the Empire became increasingly insistent, and in fact during the first three centuries of the Christian era the Empire was being slowly barbarized. Barbarians were taken into the army, settled on the lands, and on occasion held administrative posts. The gradual barbarization quickened to avalanche proportions from the fourth century. The East and the West were hit alike during the fourth, fifth, and sixth centuries, and durable success was achieved by the Germanic invaders in the West. The result was that the Germanic tribes as nations seized the major parts of the Western half of the Empire. And, although they paid lip service to the Empire at times, they achieved virtual political control. Principal success was had by the Visigoths in Spain, the Ostrogoths and later the Lombards in Italy, and the Franks and other tribes in France and parts of Germany. Just as the Barbarians felt the influence of and submitted to the Roman forms of law, administration, tenure, and so on, so in the cultural field they submitted to Latin literature. The scientific literature, we have seen, was at best mediocre in comparison with the great products of Greek scientific learning.

The Latin science of the early Middle Ages was dependent either directly or indirectly on the following sources:

(1) The earlier Roman science itself, a feeble offshoot of

Hellenic learning for which men of letters (Seneca and Pliny) must stand as representatives.

(2) The writings of the Latin fathers together with the writings of the Greek fathers that had been translated into Latin.

(3) A handful of didactic philosophical works of late antiquity, such as those of Chalcidius, Macrobius, Martianus Capella, and, of course, Boethius and Cassiodorus.

(4) The scattered Latin translations and paraphrases of Greek scientific works. This last source was never extensive and for the most part excludes the greatest of the Greek scientific writings—namely, the great mathematical works of Archimedes, Apollonius, Hero, Diophantus, etc. Also excluded were the most important works of Greek astronomy and mechanics.

We have spoken about the principal products of early Roman science as well as about some of the Christian writers of late antiquity. Now it remains to mention the didactic and philosophical works in Latin of the late antique period, works that were to influence medieval science. Our last principal source for the later Latin science, the direct translations, we shall comment upon later.

The first of the three didactic authors of the fourth and fifth centuries who were fairly popular in the early Middle Ages was Chalcidius, who flourished in the first part of the fourth century. Chalcidius is important to us because of (1) his incomplete translation of the *Timaeus* of Plato, the dialogue that concerned itself with natural philosophy, and (2) a commentary on the same work.

It was apparently Chalcidius' translation rather than an earlier one by Cicero that was used in the early Middle Ages. Thus Chalcidius provided the only *direct* knowledge of Plato's physical ideas to the natural philosophers of the early Middle Ages.

Although his basic opinions are naturally Neo-Platonic, when, for example, he describes the stages of gods, Chalcidius is not unaffected by the ideas of Aristotle. Sometimes he quotes or paraphrases the Aristotelian positions, either to dispute them or to accept them in a modified form. But so long as he passes the opinion on to later writers, his acceptance or denial of it is unimportant. He describes briefly the astronomical theory of concentric spheres; he paraphrases Aristotle's account of the soul contained in the first chapter of the book *On the Soul.* He paraphrases the Aristotelian theory of prime matter, and he presents an Aristotelian view of the transmutation of elements by reason of common qualities. His proof of the spherical form of the universe and the earth in great part stems from Aristotle, as does his distinction of the sublunar and celestial worlds. Similarly, his observations on the pre-Socratic natural philosophers are drawn from Aristotle, as well as his distinction of *possibile* from *contingens* and from *necessarium*—that is to say, the distinction of the possible from the contingent and from the necessary.

Chalcidius also transmits some of the physical opinions of the Stoic philosophers. He advances the concept of a periodic re-creation of the universe and the concept of the sowing in the world by God as Reason of rational seeds, or "semina." For him, these seeds are the fundamental principle of things procreated according to the law of nature (*lex nature*).

The most interesting part of Chalcidius' commentary occurs in the astronomical section, where he describes (probably erroneously) Heraclides' theory of the revolution of Mercury and Venus about the sun. Like Theon of Smyrna before him, Chalcidius grafts the theory of epicycles onto Heraclides' view. Thus, for him, Mercury and Venus do not revolve about the sun as the center, as they had for Heraclides; rather, the sun itself revolves in a small epicycle, and Mercury and Venus revolve in different and larger epicycles about the same center as the epicycle of the sun.

In describing the motion of Venus, and he probably assumes a similar motion for Mercury, Chalcidius says: "Lastly Heraclides Ponticus, when describing the circle of Lucifer as well as that of the sun, and giving the two circles one center and one middle, showed how Lucifer is sometimes above and sometimes below the sun" (Chapter 110, translation of T. L. Heath, *Aristarchus,* p. 256).

A similar account of Heraclides' doctrine is attributed to the second of the late antique Latin Neo-Platonic writers,

Macrobius, who flourished about 400. This account occurs in Book I, Chapter 19, of his *Commentary on Cicero's Dream of Scipio*. In fact, this is much more a commentary on the *Timaeus* of Plato than on the *Republic* of Cicero. The astronomical passage runs as follows: "The circle (epicycle) which the sun traverses is lower than and encircled by the circle of Mercury, which in turn is encircled by the superior circle of Venus." Actually, according to William Stahl, "Macrobius gives no indication of comprehending this theory in his vague allusion to it; rather he is firmly defending a fixed Platonic order of the planets—a view that is completely incompatible with Heraclides' theory." (Stahl, in *Scripta Mathematica*, Vol. XXIII, 1957, p. 180.)

In addition to the astronomical reference there is a considerable elaboration of Pythagorean number mysticism (see Book I, Chaps. 5 and 6). In rather rambling chapters on the celestial bodies, Macrobius reports the figure of 252,000 stades for the circumference of the earth. This figure is reputed to be a correction made by Eratosthenes of his original value of 250,000 stades. In these chapters Macrobius also describes the doctrine of the harmony of the spheres attributed to the Pythagoreans, the doctrine that the number and movement of the celestial spheres is such as to produce several distinct but harmonious tones. Macrobius declares further that certain Platonists hold that the distances of the planets from one another are such as to produce harmonious accord.

One modern student of Macrobius has detected direct usages of Macrobius' work in the early Middle Ages, enough to prove conclusively the widespread influence of this Neo-Platonist. Authors, such as Boethius and Isidore, whose works we shall treat presently and who themselves became very popular indeed, were also familiar with and employed Macrobius' commentary.

A third pagan author of late antiquity who had influence on the Latin science of the early Middle Ages was Martianus Capella, who flourished in the fourth or fifth century. He compiled a treatise on the seven liberal arts entitled *The Nuptials of Mercury and Philology*. The arts treated were: grammar, rhetoric, dialectics, arithmetic, music, geometry, and astronomy. The grouping of most of these subjects together as liberal arts probably goes back at least to the first century B.C., but the tendency to set the arts at the fixed number of seven hardens in canonical fashion in about the fourth century after Christ. As a result of Martianus' treatise and the adoption of seven liberal arts as the basic set of disciplines by Cassiodorus and Isidore, these arts are transmitted as the standard curriculum of the early Middle Ages. It is well known that the first three were called the *trivium* and that the mathematical subjects, grouped together by the early Pythagoreans, formed the *quadrivium*.

Martianus' work is cast into allegorical and mythological form. It deals with the wooing and marriage of Philology, "the most learned maiden," by the god Mercury. The seven liberal arts are attending maidens, and each makes a speech describing herself. The quadrivium, which is of interest to the historian of science, occupies Books VI-IX. In the first part of the treatment of geometry (Book VI) there is also contained some geography which is drawn largely from Pliny. It states that the shape of the earth is not a plane nor concave but is "rotund and global." The part on geometry proper is very short and is based at least indirectly on Euclid. It contains only definitions of various geometric magnitudes as well as postulates and axioms of the first book of Euclid. The arithmetic section has material similar to that of Macrobius. It depends on Nicomachus and thus, of course, indirectly on Euclid. The section on astronomy is drawn at least in part from the Roman encyclopedist Varro. It is of interest for repetition once again of Heraclides' idea of the revolution of Mercury and Venus about the sun. But Martianus' account is more accurate than those of Chalcidius and Macrobius, for it does not introduce falsely the idea of epicycles into Heraclides' account. He tells us that Venus and Mercury do not encircle the earth but rather "circle around the sun in a free circuit. Then it is the sun which they take as the center of their circle." Martianus' work was popular in the early Middle Ages, as its appearance in the catalogues of monastic libraries would indicate. Four of the principal manuscripts date from the tenth century, and one of the most popular of the early medieval authors, John the Scot, in the ninth century made the *Nuptials* the object of a commentary. There were in addition at least two other commentaries on it, dating from the early medieval period.

II

As we come to the end of the antique period and the beginning of the Middle Ages, we encounter two writers of the sixth century who occupy significant positions in the transmission of ancient learning to the Middle Ages. These transitional figures were Boethius and Cassiodorus. Neither of these Roman statesmen-authors can be classified as a scientist, but each played an important role in the history of science which we cannot neglect.

Boethius (*ca.* 480-524) was a Roman by birth and held a series of important political offices under Theodoric the Ostrogoth, whose favor he finally lost. The unhappy imprisonment and execution of this distinguished Roman is a well-known story. Students of philosophy from his day to this have read with admiration his famous *Consolation of Philosophy*, written while he was in prison. But more significant for the history of science were: (1) his translations, commentaries, and extensions of Aristotelian logical writings and (2) his manuals for the subjects of the mathematical quadrivium. Boethius in his mathematical handbooks continued in Latin the tendency already apparent in Martianus Capella's *Nuptials*. Thus he attempted to comprehend the whole extent of a subject in a short compendious work. It is of incidental intelligence to note that Boethius appears to have been the first to use the Latin term *quadrivium* to embrace the four mathematical subjects long associated together by the Pythagoreans. He speaks of the pursuit of the quadrivium as the way to perfection in the discipline of philosophy. And it is through the quadrivium that one is led from the senses to the surer things of intelligence.

The first of Boethius' handbooks, that on arithmetic, was based almost completely on Nicomachus' *Introduction to Arithmetic* (see Chap. 9 above); in fact, it has often been called a translation. It should be observed, however, that

Boethius compressed much of the material found in the original work and occasionally added remarks. Of some interest is his statement (I, 1) that "nothing which is infinite can be found in science nor be comprehended in science." Also noteworthy is his definition of atoms (II, 4) and his comparison of the three principal types of proportions to the political forms of oligarchy, aristocracy, and democracy. The significance of this arithmetical work lies not in its originality but in the fact that it was probably the principal source of arithmetic in the Latin West in the early Middle Ages.

The second of Boethius' mathematical handbooks, still extant, deals with music. Like the *Arithmetic* this treatise was based on anterior Greek treatises, principally those of Ptolemy and Nicomachus. One of its most interesting passages, which in theory goes back at least to Euclid, connects pitch with frequency of vibration: ". . . the same string, if it is tightened further, gives a higher pitched sound; if it is loosened, a sound of lower pitch. For when it is tighter it renders a swifter impulse and returns more quickly, striking the air more frequently and more densely." We have already quoted Boethius' exposition of a "wave theory" of sound in Chapter 6. Like *Arithmetic*, the *Music* of Boethius introduced the fundamentals of a Greek discipline to the early Middle Ages. The other two mathematical handbooks of Boethius, on geometry and astronomy, are lost. There is an extant work on geometry in at least two versions that goes under the name of Boethius, but this is generally considered to be spurious. It purports to be or to include a translation of Euclid, but what a lamentable rendering it is! For the most part the propositions are given without proof and only certain selected propositions are included. It is possible that this extant version was based on the original, a more inclusive translation of Euclid, for there seem to be remaining fragments of a genuine translation of the *Elements* of Euclid made by Boethius. Of considerable interest in one version of this spurious treatise is the description of Hindu numerals. It is generally considered, however, that this reference to Hindu numerals is a later interpolation resulting, perhaps, from contact with Arabs in Spain in the tenth and eleventh centuries. We shall return to the question of the origin of these numerals in a later volume.

That Boethius also composed a handbook of astronomy is inferred from a letter written to Boethius by Cassiodorus. In this letter, Cassiodorus praises Boethius for having made available in Latin the music of Pythagoras and the astronomy of Ptolemy, as well as the arithmetic of Nicomachus and the geometry of Euclid. The same letter tells us of still another translation by Boethius not now extant; it is of some work of Archimedes, possibly the *On the Equilibrium of Planes*. For Cassiodorus tells Boethius that he has restored to the Sicilians in Latin form their mechanician, Archimedes. We also have a reference by Boethius himself to a work that he composed on physics, but it too is lost.

Boethius' efforts to make the logical works of Aristotle and kindred logical treatises of other Greek authors available in Latin was a part of a greater scheme: to translate Aristotle and Plato into Latin. Unfortunately he died before he accomplished this task. Still, his translations of the logical works of Aristotle were of great importance for the early Middle Ages. The translations that he made are as follows: (1) the *Categories* of Aristotle, (2) Aristotle's *On Interpretations*, (3) two commentaries on Aristotle's *On Interpretations*, (4) a translation of the Neo-Platonist Porphyry's introduction to the *Categories*, (5) two commentaries on Porphyry, (6) a commentary on the *Topics* of Cicero, and (7) possible translations of the *Topics*, *Prior Analytics*, and *Posterior Analytics* of Aristotle. If the translations of these last three works were made they were probably not used extensively until the twelfth century. In conjunction with these various translations of logical works we should note also that Boethius composed a series of independent works on logic that drew on the logical achievements of the Stoics. To Boethius, then, the early Middle Ages owed most of what it knew about logic.

A word must be said on the theological tracts attributed to Boethius. At least four of them are often accepted as genuine. Some students have seen in these tracts the germ of the method followed in the schools of the high and late Middle

Ages, the so-called "scholastic method." Although this may be an overenthusiastic evaluation of these short treatises, there is no question that Boethius attempts in them to use Aristotelian logic and philosophy for the exposition of theological problems. Similarly, his general division of knowledge, which separates physics, mathematics, and theology, is Aristotelian. One remaining point of interest in these theological tracts is a note he makes that he will follow the example of mathematics and similar disciplines in laying down terms and rules on the basis of which he will proceed. And in one such treatise he defines an axiom in the manner of Aristotle and Euclid. He says "a common conception (axiom) is a statement which anyone (immediately) affirms upon hearing it." He goes on to say that such common opinions are of two kinds: One is generally acceptable to all people; for example, "if equals are taken from equals the remainders are equal." And the second is a conception generally known to the learned only; his example of such a conception is, "Things which are incorporeal cannot occupy space."

One final word might be said of Boethius' last work, his *Consolation of Philosophy*. It reflects a wide variety of philosophical opinion, Platonic and Aristotelian alike. Dame Philosophy, of grave countenance and glistening clear eyes, visits Boethius in his prison and leads him by discussion from his despair to contentment in reason and virtue. In the course of her discussion in Book III she states that she will employ the geometrician's type of argumentation to deduce *porismata* (deductions) or *corollaria* from demonstrated propositions. The ultimate object of this argumentation is to prove that the substance of God consists in nothing but goodness. In examining goodness, Boethius notes as a fundamental rule of nature that every living creature labors to preserve his health and to escape injury and death. This is true of plants as well as animals, and even of those things thought to be without life. "For why does levity lift flames and heaviness depress things of earth—only because these places and movements are fitting for them." Goodness is, in fact, the purpose of the universe. The teleological point of view is as strongly imbedded in this work of Boethius as it is in the works of Aristotle. The wide popularity of the *Consolation* in the Middle Ages is well known. It was rendered into Anglo-Saxon by King Alfred in the ninth century, into old High German in the eleventh century, into Middle English by Chaucer, and so on.

Of less intellectual power and scientific knowledge was Cassiodorus (*ca.* 488-575). Like Boethius, he was a Roman who held many important positions under the Ostrogothic kings. His overwhelming importance for the history of learning lies in his prescription to the monks of the monasteries that he founded to copy the monuments of divine and human letters. This prescription occurs in his *Introduction to Divine and Human Readings,* written sometime after 551. He was not the first to urge such monastic labor, but because of its detailed advice and the extent of such advice, his was the most influential call to the preservation of manuscripts. It is true that we are told by Cassiodorus to cultivate the liberal arts in order to increase our capacity for understanding sacred literature, but fortunately he took the question of preserving profane as well as religious literature most seriously. We know that he collected manuscripts to serve as the models for copying. He details carefully what rules of spelling are to be followed in copying and what errors are to be avoided. Besides listing some of the important works of science that he thinks should be preserved, he gives us some indication of his own understanding of Greek learning. For example, he describes the basic Aristotelian classification of the sciences, first in diagrammatic fashion and then in more detail. The diagrammatic classification follows:

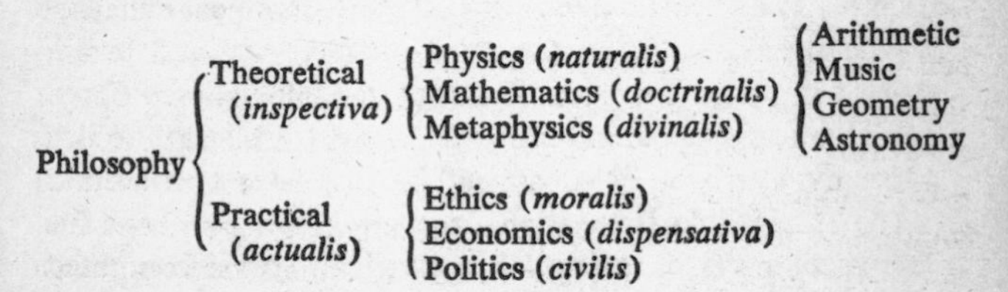

The mathematical sections in the book are more elementary and brief, with definitions predominating. Like that of Boethius, the influence of Cassiodorus in the early Middle

Ages was extensive. His principal work was often used and cited, and the extant manuscripts dating as early as the ninth and tenth centuries are numerous.

III

Before going deeper into the Latin science of the early Middle Ages we can profitably scan and summarize the translations of Greek scientific works into Latin in antiquity. As we have mentioned before, there is a distinct absence of translations of the best mathematical and mechanical works. The predilection for medical works should not escape attention. It will also be noted that, so far as our knowledge of these translations is concerned, they seem to stem largely from the later antique period. In listing the translations below we have maintained a rough chronological order. This list does not pretend to be absolutely complete but does mention the most important scientific translations or direct paraphrases of Greek scientific writings made in antiquity.

(1) Celsus' paraphrase or translation of an unknown Greek medical work (see Chap. 8, Sect. V).

(2) Vitruvius' paraphrase of Ctesibius and other mechanical authors (see Chap. 8, Sect. IV).

(3) The possible translation of the *Timaeus* of Plato by Cicero; but it was the later incomplete translation of Chalcidius that was used in the Middle Ages (see Chap. 11, Sect. I).

(4) One translation of the *Phaenomena* of Aratus by Cicero. The *Phaenomena* was an astronomical-meteorological poem written in the third century B.C. and based, at least in part, on a work or works of Eudoxus (see Chap. 7). There was apparently also an early translation of this poem by Varro of Atax, as well as two later translations.

(5) A possible translation or compilation by Pompeius Trogus of parts of Aristotle's zoological works; composed under Augustus and used by Pliny in the first century.

(6) Possibly a work on plants based on Theophrastus, by the same Pompeius.

(7) The translations of Nicomachus' *Introduction to Arithmetic* and his *Harmonics* by Apuleius in the second century (see Chap. 9, Sect. I).

(8) A translation by Apuleius of the pseudo-Aristotelian *On the Universe* (*De mundo*).

(9) The sundry paraphrases of Greek works in natural history that appear in Pliny's *Natural History* (see Chap. 8, Sect. VI).

(10) An anonymous translation of Books XI-XIII of Euclid's *Elements* in about the fourth century, of which only fragments are extant.

(11) Anonymous translation of about the fifth century of several of Hippocrates' works, including the following: *The Aphorisms; Airs, Waters and Places; On Diet; On Sevens; Prognostic.*

(12) Anonymous translations of about the fifth century of the following of Galen's works: *Therapeutics; On Fevers; On Medicinal Simples.*

(13) Miscellaneous and partial translations and paraphrases of other Greek medical authors, such as Rufus of Ephesus, Oribasius, and Soranus, e.g., the translations from Soranus' *Gynaecia, Acute Diseases,* and *Chronic Diseases* by Caelius Aurelianus (possibly in the fifth cent. A.D.). (See Drabkin's translations.)

(14) An anonymous translation, entitled *Exposition of the Whole Universe,* of an unknown Greek original, done after 412.

(15) Probably three different translations of the work *On Medicine* of Dioscorides.

(16) A medical paraphrase of Galen and others done by Cassius Felix about 440, entitled *On Medicine.*

(17) The translations and paraphrases of several Greek geographical works, such as the *Descriptio orbisterrae* of Dionysius of Alexandria.

(18) The numerous translations and elaborations of Aristotelian logic of Boethius (see Chap. 11, Sect. II).

(19) A possible translation of one of the mechanical works of Archimedes by Boethius (see Chap. 11, Sect. II).

This work apparently did not survive even into the early Middle Ages. Which, if any, of Archimedes' works were translated is not known.

(20) A translation or paraphrase by Boethius of some astronomical work of Ptolemy, which was also lost at an early date.

(21) The possible translation of Euclid's *Elements* by Boethius (see Chap. 11, Sect. II).

(22) Boethius' paraphrase and translation of Nicomachus' *Introduction to Arithmetic*.

(23) A possible epitome of Aristotle's *Physics* by Boethius. If it existed at all, it does not appear to have survived into the early Middle Ages.

(24) Several paraphrases and translations of Greek works on music, including that of Nicomachus, already mentioned, and that of Gaudentius, the latter being translated in the sixth century by one Mucianus.

(25) An early medieval translation of the medical handbook (*Therapeuticon*) of Alexander of Tralles (see Chap. 13, Sect. III).

After closely examining this list and considering it alongside the much more extensive number of works translated from Greek into Arabic in the ninth and tenth centuries, one will readily appreciate the fact that in comparison with the scientist of Islam the early medieval natural philosopher in the Latin West had only meager scraps of the Hellenistic corpus of scientific writing.

Chapter 12

Latin Science in the Early Middle Ages

I

BOETHIUS, Cassiodorus, and their predecessors in the last centuries of the Western half of the Empire set the form and to a large extent the content of early medieval natural philosophy. Compendious encyclopedias and epitomes containing only the simplest scientific propositions became the rule. Of the authors of these encyclopedias and epitomes we should first mention Isidore of Seville, who succeeded his brother as Bishop of Seville about 600 and died in 636. Less than a century separates St. Isidore from Boethius. But the gulf in scientific and philosophical learning that separates the two men is profound. Boethius was directly bound up with Greek learning, and his knowledge of the Greek language and some of its finest products was sure and sensitive; it was no accident that Boethius was able to add something original to the growth of propositional logic. Isidore, on the other hand, seems to have had but a superficial knowledge of Greek learning; and his understanding of the subtleties of Greek science is either elementary or completely lacking.

In our investigation of the science of the early Middle Ages we are particularly interested in two of Isidore's works: *On the Nature of Things* and the *Etymologies*. The first of these is a short treatise including a diversity of topics, such as chronology, elementary astronomy, cosmology, meteorology, and geography. One confused chapter of this work (Chap. 10) appears to be an attempt to apply the Greek theory of climatic zones to a flat world. If this is so, then Isidore stood practically alone among the natural philosophers of the Middle Ages in his belief in a flat earth. It would appear to be more likely, however, that Isidore was simply doing a rather poor job of explaining and describing a Greek theory which he did not completely understand, and that he did not actually believe in a flat earth. For in the first place he describes the zones or climes more correctly elsewhere. In the second place, he speaks of the circumference of the earth, giving its value as 180,000 stades. If he understood any of the sources from which he might have taken this figure, he could hardly have escaped the conclusion that the earth was spherical in shape. The value of 180,000 stades was one value reported by Strabo the Geographer and was also given by Ptolemy.

In discussing the nature of the elements Isidore recounts the Aristotelian doctrines generally known and repeated by the Latin authors of late antiquity. He also mentions the celestial fifth element, or ether. He relates the elements to the constitution of man and to the four bodily humors. Meteorological doctrines are explained in terms of the elements. Painfully unsophisticated is his notion of spring as composed of moisture and heat, summer of fire and dryness, and so on. Similarly, the four cardinal points are related to the seasons and to the elements in an entirely unsatisfactory manner.

About Isidore's *Etymologies* little need be said. Written in twenty books, it runs through the seven liberal arts, medicine, law, chronology, religion, man, natural history, mineralogy, geography, and a miscellany of other topics. Rarely is any topic discussed with maturity. The treatment is always brief, in dictionary form, and reveals little evidence of independent observation. The etymological derivations from which the work takes its title are often ingenious, if in error.

The section on the four mathematical subjects is like the treatment of Cassiodorus, of an abbreviated, definitional character, without interest to the historian of mathematics. Much comes through the late Latin authors from Nicomachus. The long book on animals contains a wide variety of species with both real and fabulous habits and characteristics. Solinus and Pliny appear to be the sources for this section. The division is made between domestic and wild animals. The two groups are described. Then follow descriptions of insects, serpents, worms, fish, birds, and minute flying animals such as the bee. Thorndike judges that there is proportionally less superstitious matter in Isidore's works than in Pliny's. Still Isidore cites numerous cases of mythical monsters and portentous births. From the fact that he uses the phrase "this has been proved by numerous experiences" we must not conclude that Isidore himself was a mature naturalist, seeking to confirm the facts of natural history by careful observation in the manner of Aristotle and Theophrastus. We would be closer to the mark in affirming Thorndike's judgment that the chief importance of the *Etymologies* consists "in showing how scanty was the knowledge of the early middle ages." It should be noted finally that the medieval authors after Isidore tended to follow the arrangement found in the *Etymologies* rather than that of Pliny's *Natural History*. In fact, the *Etymologies* served as a kind of bridge between the Roman authors of antiquity and the early medieval writings.

Although Isidore surpasses his contemporaries in the extent of the subject matter he covers, his individual treatment of a single subject tends to be more cursory than corresponding discussions in the few treatises on separate scientific topics which survive from this day. We can note, as an example, the treatise entitled *On the Course of the Stars,* composed by Gregory, Bishop of Tours, from 573 to 594. Educated in Gaul, Gregory appears to have been acquainted with the *Nuptials* of Martianus Capella. His fame rests largely on his well-known *History of the Franks.* As for his short astronomical tract, to which he refers in one place as *De cursibus ecclesiasticis,* we are told that its purpose is not to teach *mathesis* (astronomy) or to warn about the future but rather to provide information about the nightly course of the constellations so that the clergy might know when are the proper hours for their night offices in different months of the year.

The little work begins with a description of the seven manmade wonders of the world: Noah's Ark, the gardens of Babylon, the temple of Solomon, the tomb of a Persian king (the Mausoleum?), the statue of the Colossus at Rhodes, the theater at Heraclea, and the Pharus or lighthouse at Alexandria. But there are also wonders of God in the universe, natural wonders repeated every day or every year. Examples are the tidal movements, the growth of plants, the phenix, Mount Etna, the hot springs near Grenoble, the sun's course, and the phases of the moon. From this introduction Gregory proceeds to a description of some of the stars and constellations and a statement of their varying times of visibility throughout the year. As the text demands and the eighth-century Bamberger manuscript of the work shows, the descriptions were accompanied by pictorial diagrams. Let us take as an example the beginning of the brief entry for the Big Bear:

URSA MAJOR

Chapter 33. Of these stars, which rustics call the Wagon, what can we say, since they do not rise and set like the other stars. However little we know about them, we should not remain silent. . . . (*De cursu stellarum ratio*, translated from the edition of Br. Krusch, Hanover, 1885.)

FIG. 24. Ursa Major, according to an eighth-century Bamberger manuscript.

Another work, which purported to be a specialized work on geography translated by Saint Jerome from the Greek of Ethicus Cosmographicus, was in reality a work of about the eighth century which depended heavily on Isidore. Of somewhat more interest is the geographical survey of an anonymous writer of Ravenna dating from about the same period, which is an itinerary or a list of place names; in part, at least, it goes back to the earlier survey called the Peutinger Table, the accuracy of which has recently been supported. Other compilations of earlier geographical statements scarcely include enough original material to merit separate treatment.

II

Among the succeeding authors who drew on the work of Isidore but who were not without individual importance was the able English church historian Bede (*ca.* 673-735). Bede's work is a monument to the accelerated intellectual activity in the late seventh and eighth centuries. Some of the credit for this English activity must be given to the Greek monk Theodore of Tarsus, who was elevated to the Archbishopric of Canterbury in 669, and to his contemporary Hadrian, formerly an abbot of a monastery near Monte Cassino, in Italy, and later of the abbey of St. Peter and St. Paul at Canterbury. As Bede tells us in his *Ecclesiastical History of the English Nation,* both Theodore and Hadrian were well read in sacred and profane literature. In addition to teaching sacred literature, they taught astronomy and arithmetic. "A testimony of their influence is that there are still living at this day some of their scholars who are as well versed in the Greek as in the Latin tongue."

To the influence of Theodore and Hadrian on the English development we must add the undoubted influence of the Irish monastic education. It was evident particularly in the evolution of a quasi-scientific genre to which Bede was to make notable contributions—namely, the *computus,* a collection of materials to be used in the construction and understanding of ecclesiastical calendars, i.e., for the preparation and understanding of Easter tables. One of the foremost students of the *computus,* Jones, has given the following characterization of the genre:

> The manuscripts of this type written from, say, the eighth to the twelfth centuries, number several hundred, and when they are properly organized they can reveal to us fruitful information about the scientific life of those ages. . . . These *computi* consist of extracts or complete tracts, often either anonymous or attributed to the wrong author, Easter-tables, a yearly calendar, lists of calculations, accessory tables for help in calculation, computistical verses for memorizing, dialogues for school catechism, and multiplication tables. These combinations very shortly attracted to themselves works on arithmetic, astronomy, geography, chemistry, and medicine. (C. W. Jones, *Bedae Opera de temporibus,* Cambridge, Mass., 1943, pp. 75-76.)

In one computus (Migne, *Patrologia Latina,* Vol. 129, 1274-

1372), for example, there are some 156 computistical items ranging in length from a fraction of a column to several columns. It contains numerous repetitious items, since it was drawn from many sources and put together loosely. One of the chapters (Chap. 48) summarizes some of the objectives of the *computus*:

> Hence by the first reckoning you ought to inquire concerning the 19 year (lunar) cycle. By the second, the disposition of the moon through the calendar of months. By the third, how the moon has a number in its month. By the fourth, what sort of day and moon is the Pascal solemnity. By the fifth, how Easter is to be computed. By the sixth, the ascension and descending of the sun. By the seventh, the response to those who are in discord over the computation of Easter.

As to the scientific content of the *computus*, let us limit our citation to a single computistical formula for finding out the age of the moon on a particular day.

> If you wish to know on this or that day what the number of the moon is (i.e., how many days after new moon), reckon the number of days from the Kalends of January up to the day you are inquiring about. And when you know this, add the age of the moon on the Kalends of January. Divide the total by 58. If more than 30 remains subtract 30, and the remainder is the moon of the required day. . . . If you wish to know the moon of the Kalends of January, take the cycle of the present year (i.e., the number of the year in the current 19 year lunar cycle). Multiply it by eleven. Divide this by 30. Add (ordinarily one). The result is the age of the moon on the Kalends of January.

The medieval *computus* as a corpus of items seems to have had its beginnings in Spain. From there it traveled to Ireland and England and then back to the Continent, where it played an important part in Carolingian scientific education. Bede, standing solidly in the computistical tradition, applies critical methods to that tradition and leaves the *computus* in more understandable form.

Bede's interest in the *computus* was demonstrated principally by two works: *On times* (*De temporibus*) and *On the Reckoning of Times* (*De temporum ratione*). The former was composed in 703, the latter twenty-two years later. The second of the treatises is an elaboration of the first. Neither work is of high scientific caliber when compared with Greek science, yet there is a certain internal consistency and clarity that are missing in the works of Isidore. Bede appears to have been careful to understand something before he borrowed and used it.

The most noteworthy part of Bede's *De temporum ratione* deals with the tides. In Chapter 29 he observes the differences between the stronger and weaker tides (*malinae* and *ledones*, "living" waters and "dead" waters). He also notes that prevailing or contrary winds can advance or retard the hours of flux and reflux and even the days of the strong and weak tides. Most important, he advances what is essentially the concept of "establishment of the port"—i.e., that the tides can be established as roughly constant for a given port, or to put it somewhat more scientifically, that the interval between the moon's meridian passage and the high water that follows is roughly constant for a particular port on the globe, but varies in length from port to port. He supported this idea with observations. It is believed that Bede's remarks on the tides stimulated the construction of tide tables based on the nineteen-year lunar cycle. At any rate, such tables appear in the later *computus* (see Fig. 27).

It should be pointed out as of scientific interest that the computistical material did much to stimulate the use of tables and diagrams. Although the tables abound in error and, as Bede points out, encourage the lazy, their construction stimulated at least elementary scientific ratiocination. We have included here a most interesting diagram that appears in numerous astronomical *computus* manuscripts, some of which date back to the ninth century (see Fig. 25). It attempts to plot the paths of the sun, the moon, and the planets through the zodiac in accordance with Pliny's description of their movements. Notice that a graph has been used with twelve divisions of latitude and thirty divisions of longitude. Such divisions probably arise from a statement made by Martianus Capella. It is possible but by no means proved that such astronomical graphing influenced the application of the graphing technique to problems of local motion and qualitative changes in the fourteenth century.

This diagram should be compared with a diagram suppos-

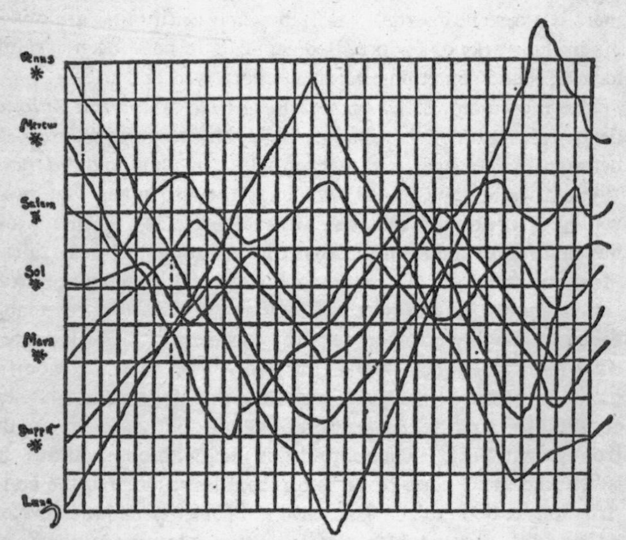

FIG. 25. An early medieval graphing of planetary movements (according to Günther via Lattin; see *Isis,* Vol. 38, 1948, p. 216).

edly representing the apsidal lines of the sun and the planets (see Fig. 26). Among the other *figurae* that appear in the astronomical-computistical writings is the well-known *rota,* or wheel, connecting the tides with the moon's phases (see Fig. 27). To give a final example of astronomical illustration in the early Middle Ages we can point to a diagram without textual elaboration appearing in a ninth-century manuscript; it represents the system of Heraclides of Pontus, with Mercury and

Venus encircling the sun rather than the earth. Included with the main diagram is a close-up showing Venus with an oval (or elliptical?) orbit, whereas Mercury has a circular one.

Much less interesting than the computistical writing to which Bede contributed was his primer *On the Nature of Things.*

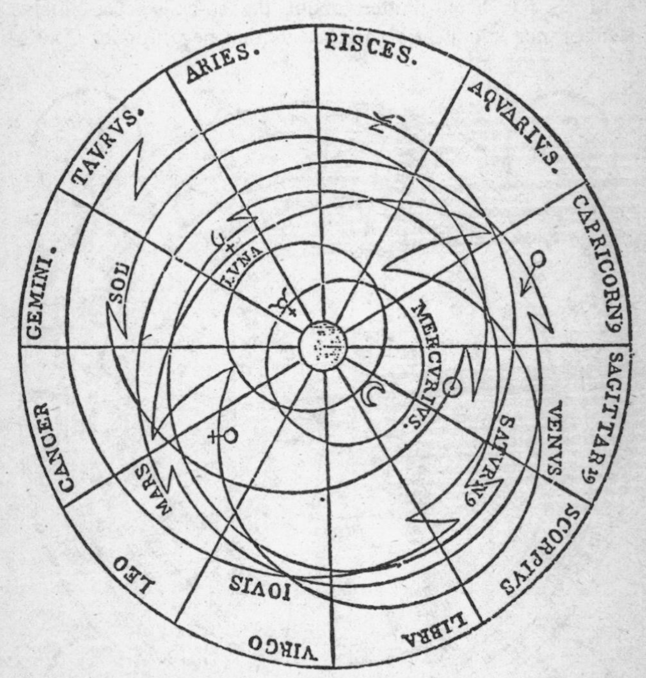

FIG. 26. An early medieval diagram purported to represent the aspses of the planets. (Migne, *PL*, 90, c. 227.)

Although it followed the model and often the words of Isidore's treatise of the same name, it is generally more lucid and depends in part on Pliny's *Natural History*. It was probably responsible for an increased interest in Pliny in the Carolingian period immediately following, demonstrated by the fairly wide circulation of an encyclopedia of excerpts from Pliny. The ambiguity on the shape of the earth in Isidore's treatise is gone in this work of Bede. Bede clearly supports the spherical earth of Greek science.

III

Much has been written about the so-called Carolingian Renaissance in education and letters that began under Charle-

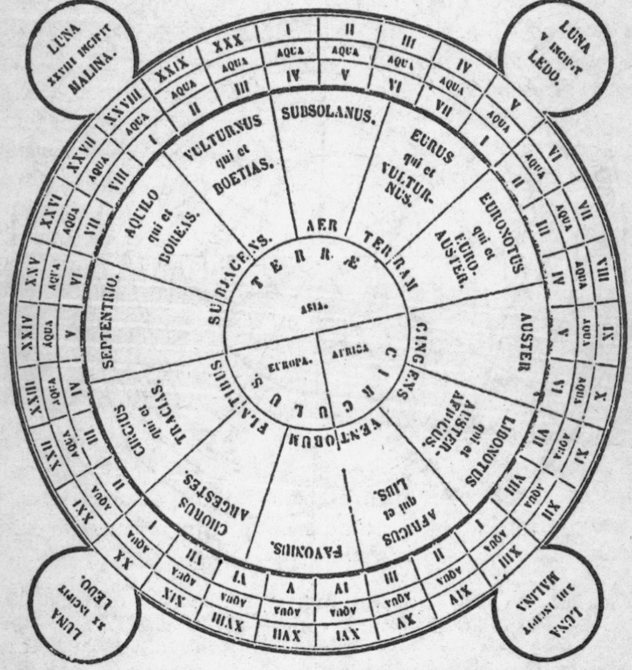

FIG. 27. An early Lunar-tide wheel illustrating the medieval computus. (Migne, *PL*, 90, c. 259.)

magne (sole ruler of Frankland from 772, emperor from 800) and continued under his successors. Scholars migrated to France from England, from Italy, and even from what was left of Visigothic Spain after the Arabic incursions of the peninsula. Thus the English scholar Alcuin of York played an important role in the Palace School of Charlemagne. The works written by Alcuin apparently in connection with this school were dialogue in form and, where scientific topics are treated, most elementary. Alcuin's greatest achievement seems to have been the inauguration of a tradition of education in northern France that was to have important results in the ninth and tenth centuries.

But educational reform was not accompanied by the collapse of all opposition to the liberal arts as pagan-inspired. One critic, after lambasting the liberal arts one by one, comes finally to the astronomers, "who have wished to soar heavenwards, but so conspicuously have they failed to mount thither by their ideas that rather they have placed earthly reason in heaven and heavenly reason in earth. For the while they have set rams and bulls, scorpions and crabs, lions and bears, she-goats and fishes in the realm of the sky, they have done naught but raise up earthly things into the sky." (M. L. W. Laistner, *Thought and Letters in Western Europe* 500-900, London, 1931, p. 168.)

Of the subjects comprising the mathematical quadrivium, geometry fares least well in the early stages of the Carolingian reforms. There is no evidence of any geometrical knowledge beyond the elementary handbook attributed to Boethius and the few statements in Cassiodorus and the encyclopedists. It was not until the time of Gerbert (*ca.* 940-1003) that geometrical studies began to pick up. Of this activity of Gerbert we shall speak in another volume. Arithmetic and astronomy centered in the passages in the encyclopedias, the Boethian handbooks, and the *computus* corpus.

A number of authors of the Carolingian age would deserve mention in a more general study of thought and learning in the eighth and ninth centuries. But they have little importance for a study of the science of the period. However, the works of one of the ninth-century authors touch on scientific matters; these are the writings of John Eriugina, known as the Scot (*ca.* 800-877). His knowledge of Greek was extensive. Furthermore, it was not merely passive knowledge, for he translated a Neo-Platonic work. His own *Division of Nature* also shows Neo-Platonic influences. The only part of this work of interest to the historian of science is the astronomical section. He gives an account amplified from Martianus Capella, of how Eratosthenes found the circumference of the earth (see Chap. 6). He also gives the value of 252,000 stades for the circumference, as well as that of 180,000 stades, attempting to reconcile the two. But the whole discussion shows a rather shaky knowledge of mathematics and astronomy, if the published texts are accurately representative of John's work. But there is one interesting passage in which John appears to have extended the system of Heraclides of Pontus, so that not only are Mercury and Venus thought to be encircling the sun rather than the earth, but Jupiter and Mars as well, although some recent authors, the von Erhardts, have attempted (without success, it seems) to give this passage another interpretation. In his *Annotations to Martianus Capella* John attributes to Plato the idea that the sun is the center of *all* the planets with the exception of the earth (*Annotations*, 13-3). This is essentially the system adopted by Tycho Brahe in the sixteenth century, a system wherein all the planets revolve about the sun and the sun and its planetary satellites revolve about the earth, which is at rest in the center of the universe. We are not to infer from John's improved system that he was a practicing astronomer, but rather that he was a diligent student of past astronomical literature with a certain turn of speculative originality. To John is also attributed a translation of the *Solutions* written by one Priscianus Lydus in the sixth century. The Greek original is now lost. This work contains elementary problems of natural science of little interest to us.

In our survey of early medieval science we have not discussed technological chemistry or medicine. In both of these fields there was activity worthy of report. In each case we shall defer treatment until, in a later volume, we can discuss the entire medieval development.

Our final judgment of Western Latin science to about 900 is that it was on a considerably lower plane than was science in the Greek East. The tendency toward the encyclopedia and toward epitomizing earlier authors appears in both areas, but in the Greek East the encyclopedias are complemented by the classical texts of Greek science, as well as by original and searching commentaries on those texts. In the West the only area of learning where we find anything like sustained scientific investigation is that of the *computus*. But even there the few important evidences of scientific learning are bogged down in a mire of uncritical summary and quotation of past authors.

Chapter 13

Greek Science in the Age of Justinian

I

WE HAVE pursued the meager trickle of Latin science through the late antique and early medieval periods to about 900. Now we must backtrack to the main stream of scientific development in the East. In the sixth century, before the Greek corpus of learning in the East was taken up and translated into Syriac and Arabic, a significant burst of Greek science took place. We shall call this study of science in the sixth century that of "the age of Justinian," after the chief political architect of the period. Actually it is difficult to know how directly the remarkable activity of Justinian affected scientific pursuits.

It would seem reasonable that Justinian's enormous sacred and profane building program stimulated the study of mechanics and mathematics. In fact, two of the chief architects of the famous church of St. Sophia in Constantinople can be connected directly with the study of the best of the classical mathematicians. These architects were Anthemius of Tralles and Isidore of Miletus. A friend of both was the mathematician Eutocius, who flourished about 520. Eutocius prepared a version of the first four books of the *Conics* of Apollonius of Perga, to which he added a commentary of his own. This work Eutocius dedicated to Anthemius. It should be noted that it is the Greek text prepared by Eutocius that survives today. Of the remaining four books of the *Conics,* Books V-VII are extant only in the Arabic and Book VIII is lost. Hence we would judge that without the activity of Eutocius all contact with the original text of this great mathematical work of Apollonius would have been lost.

Similarly, we owe to Eutocius the preparation of commentaries on at least three of Archimedes' works: *On the Sphere and Cylinder, The Measurement of the Circle,* and *On the Equilibrium of Planes.* It is suspected that these were the three most popular of Archimedes' works at this time. The commentaries on the first two works were edited by Isidore of Miletus, i.e., the latter prepared a new edition of them. Eutocius' commentary on *The Measurement of the Circle* has proved most useful in attempts to reconstruct the calculating procedures used by Archimedes in his determination of pi. It seems likely that there was a fairly flourishing school of mathematicians under Isidore and that we owe to his and Eutocius' interest in Archimedes the survival in Byzantine Constantinople of Archimedean manuscripts. This survival became crucial in the ninth century, when Leon, the so-called reformer of the Byzantine University in Constantinople, collected the then existent works of Archimedes. It is apparently Leon's ninth-century collection that proved to be the source of our extant texts of most of the mathematical works of Archimedes, and only a single manuscript of Leon's collection seems to have survived into the Middle Ages. But for the survival of that manuscript (and, of course, the previous activity of Eutocius and his contemporaries), the Greek text of much of the work of antiquity's greatest mathematician might not have come down to us today.

II

Another source of mathematical and scientific activity during this period was the Neo-Platonic movement. We have already spoken of the mathematical interests of the Neo-Platonists through the time of Proclus in the fifth century (see Chap. 9, Sect. VI). In the sixth century the movement produced two of its most able exponents: John Philoponus (in his pre-Christian period) and Simplicius. It is true that the movement was rebuffed, but certainly not terminated, by the closure of the Neo-Platonic school in Athens by Justinian in 529 as a center of paganism. The flight to Persia under the protection of King Khusraw of seven of the group and their subsequent return to the Empire was certainly not the first nor the last academic flight from political persecution.

In recent years our estimate of the stature of John Philoponus as a natural philosopher has risen with the closer examination of his commentaries on the works of Aristotle. The historian of science is particularly interested in the commentary on the *Physics* of Aristotle, of which commentary four books are still extant. For in this commentary Philoponus subjected the mechanical views of Aristotle to a severe and often just criticism.

In our preliminary examination of Aristotle's views on movement (see Chap. 6) we noted that Aristotle assumed that for movement to take place both a motivating force and a resistance are required. From several passages in the *Physics* and the *Book on the Heavens* we can deduce the following simple "quantitative" laws:

(1) $S_1:S_2::F_1:F_2$ when $T_1 = T_2$ and $R_1 = R_2$, and assuming that movement occurs. S_1, S_2 are distances. T_1, T_2 are times. In "forced" movement F_1, F_2 are external forces; R_1, R_2 are weights. But in "natural" movements (such as that of falling bodies) F_1, F_2 are weights and R_1, R_2 are the densities of the media through which the weights fall.

(2) $T_2:T_1::F_1:F_2$ when $S_1 = S_2$, $R_1 = R_2$, and movement occurs.

(3) $S_1:S_2::R_2:R_1$ when $F_1 = F_2$, $T_1 = T_2$, and movement occurs.

(4) $T_1:T_2::R_1:R_2$ when $S_1 = S_2$, $F_1 = F_2$, and movement occurs.

One could, for the sake of economy, use a single modern formula to express all these cases, although, of course, Aristotle does not do this:

(5) $V \propto \dfrac{S}{T} \propto \dfrac{F}{R}$ with V as the speed.

Thus the basic dynamic formula which we might deduce from the scattered statements of Aristotle is that "speed is proportional to the ratio of the motive force to the resistance, provided that the force is sufficiently great to overcome resistance and produce movement." Now suppose that we had a natural movement in a vacuum. The density of the medium would obviously be zero and thus the movement would take place instantaneously (or, in the modern formula above, V would go to infinity as R goes to zero). Since an instantaneous movement appears to Aristotle to lead to contradictions, it is unthinkable, and hence a vacuum does not exist.

Eager to prove that movement can take place in a vacuum, John criticized the basic dynamic laws of Aristotle in the following passage:

> Weight, then, is the efficient cause of downward motion, as Aristotle himself asserts. This being so, given a distance to be traversed, I mean through a void where there is nothing to impede motion, and given that the efficient cause of the motion differs, the resultant motions will inevitably be at different speeds, even through a void. . . . Clearly, then, it is natural weights of bodies, one having a greater and another a lesser downward tendency, that cause differences in motion. For that which has a greater downward tendency divides a medium better. Now air is more effectively divided by a heavier body. To what other cause shall we ascribe this fact than that that which has greater weight has, by its own nature, a greater downward tendency, even if the motion is not through a plenum? . . .
>
> And so, if a body cuts through a medium better by reason of its greater downward tendency, then, even if there is nothing to be cut, the body will none the less retain its greater downward tendency. . . . And if bodies possess a greater or a lesser downward tendency in and of themselves, clearly they will possess this difference in themselves even if they move through a void. The same space will consequently be traversed by the heavier body in shorter time and by the lighter body in longer time, even though the space be void. The result will be due not to greater or lesser interference with the motion but to the greater or lesser downward tendency, in proportion to the natural weight of the bodies in question. . . .
>
> For if a body moves the distance of a stade through air, and the body is not at the beginning and at the end of the

stade at one and the same instant, a definite time will be required, dependent on the particular nature of the body in question, for it to travel from the beginning of the course to the end (for, as I have indicated, the body is not at both extremities at the same instant), and this would be true even if the space traversed were a void. But a certain *additional time* is required because of the interference of the medium. For the pressure of the medium and the necessity of cutting through it make motion through it more difficult.

Consequently, the thinner we conceive the air to be through which a motion takes place, the less will be the *additional time* consumed in dividing the air. . . .

If a stone moves the distance of a stade through a void, there will necessarily be a time, let us say an hour, which the body will consume in moving the given distance. But if we suppose this distance of a stade filled with water, no longer will the motion be accomplished in one hour, but a certain additional time will be necessary because of the resistance of the medium. Suppose that for the division of the water another hour is required, so that the same weight covers the distance through a void in one hour and through water in two. Now if you thin out the water, changing it into air, and if air is half as dense as water, the time which the body had consumed in dividing the water will be proportionately reduced. In the case of water the additional time was an hour. Therefore the body will move the same distance through air in an hour and a half. If, again, you make the air half as dense, the motion will be accomplished in an hour and a quarter. And if you continue indefinitely to rarefy the medium, you will decrease indefinitely the time required for the division of the medium, for example, the additional hour required in the case of water. But you will never completely eliminate this additional time, for time is indefinitely divisible. (M. Cohen and I. E. Drabkin, *A Source Book in Greek Science,* New York, 1948, pp. 217-219.[1])

[1] This quotation and the five succeeding quotations are quoted by permission of the publisher, the McGraw-Hill Book Company, Inc.; copyright, 1948.

For Philoponus, then, the fundamental and "original" determiner of movement is motive force. If there were no resistance —that is, if bodies were moving in a vacuum—a certain force would implant in a certain body a given movement, and the body would move a certain distance in a certain time (the "original" time).

$$(6)\ \frac{S}{T_o} \propto F.$$

Philoponus says that in addition to the original time there is a time due entirely to the resistance. This added time is directly proportional to the density of the medium. Thus John is saying that

$$(7)\ T_f = T_o + T_a \qquad \text{and} \qquad (8)\ T_a \propto R$$

where T_f is the total time taken to move through S when there is resistance R, T_o is the time taken to move through S when there is no resistance, and T_a is the additional time due exclusively to the resistance. By this doctrine of original time motion in a vacuum was saved. Later authors were also to interpret this passage as saying essentially that speed is proportional to the *difference* of force and resistance (rather than to the *ratio* of force to resistance, as Aristotle was interpreted as saying). But we shall discuss this extension of Philoponus by the Islamic authors in a later volume.

In the passage already quoted above Philoponus has clearly opined that in a vacuum bodies would fall in times inversely proportional to the weights. But when we come to actual fall in a resistant medium Philoponus demonstrates that the Aristotelian conclusion of the inverse proportionality of the weights and the times is clearly erroneous. Notice that Philoponus concludes that the time of fall for bodies of which the difference in weight is not too great is virtually the same. Notice further that he supports this conclusion by reference to the *experiment* of letting fall objects of different weight. This is the experiment supposedly first performed by Stevin or Galileo.

For Aristotle wrongly assumes that the ratio of the times required for motion through various media is equal to the ratio of the densities of the media. . . .

But from a consideration of the moving bodies themselves we are able to refute Aristotle's contention. For if, in the case of one and the same body moving through two different media, the ratio of the times required for the motions were equal to the ratio of the densities of the respective media, then, since differences of velocity are determined not only by the media but also by the moving bodies themselves, the following proposition would be a fair conclusion: "in the case of bodies differing in weight and moving through one and the same medium, the ratio of the times required for the motions is equal to the inverse ratio of the weights." For example, if the weight were doubled, the time would be halved. That is, if a weight of two pounds moved the distance of a stade through air in one-half hour, a weight of one pound would move the same distance in one hour. Conversely, the ratio of the weights of the bodies would have to be equal to the inverse ratio of the times required for the motions.

But this is completely erroneous, and our view may be corroborated by actual observation more effectively than by any sort of verbal argument. *For if you let fall from the same height two weights of which one is many times as heavy as the other, you will see that the ratio of the times required for the motion does not depend on the ratio of the weights, but that the difference in time is a very small one.* And so, if the difference in the weights is not considerable, that is, if one is, let us say, double the other, there will be no difference, or else an imperceptible difference, in time, though the difference in weight is by no means negligible, with one body weighing twice as much as the other.

Now, if, in the case of different weights in motion through the same medium, the ratio of the times required for the motions is not equal to the inverse ratio of the weights, and, conversely, the ratio of the weights is not equal to the inverse ratio of the times, the following proposition would surely be reasonable: "If identical bodies move through different media, like air and water, the ratio of the times required for the motions through the air and water, respectively, is not equal to the ratio of the densities of air and water, and conversely." (*Ibid.*, pp. 219-220.)

It is somewhat difficult to reconcile this experimental evidence with Philoponus' earlier opinion on the relationship of weight and speed in a vacuum; but we must suppose that the effect of resistance is such as to nullify difference in weight. A further but unlikely possibility is that Philoponus held the principal determiner of the speed of movement in a resistant medium to be specific weight rather than gross weight, and thus in the experiment above he drops different weights of the same substance. Benedetti and Galileo in his youth were to hold such an opinion.

Besides criticizing Aristotle's dynamic laws Philoponus attacked Aristotle's curious views on projectile motion. Aristotle had supposed that the air in some fashion exerts the continuing force that keeps the projectile moving when it is no longer in contact with the projector. Thus the movement of the air is giving the role of a motivating force in projectile motion. Aristotle was uncertain of the detailed mechanics involved in the air's role as propellant. But he seems to have believed either that the air accomplished its pushing force by filling a temporary vacuum formed behind the projectile as it moved forward or that the air, being more elastic and more easily excitable and having a greater facility for being moved, retains its turbulence and thereby successively pushes the projectile in the direction toward which it has been stirred. Philoponus directs a devastating appeal to experience and common sense against these views:

Let us suppose that *antiperistasis* takes place according to the first method indicated above, namely, that the air pushed forward by the arrow gets to the rear of the arrow and thus pushes it from behind. On that assumption, one would be hard put to it to say what it is (since there seems to be no counter force) that causes the air, once it has been pushed forward, to move back, that is along the sides of the arrow, and, after it reaches the rear of the arrow, to turn around once more and push the arrow forward. For, on this theory, the air in question must perform three distinct motions: it must be pushed forward by the arrow, then move back, and finally turn and proceed forward once more. Yet air is easily moved, and once set in motion travels a considerable distance. How, then, can the air, pushed by the arrow, fail to move in the direction of the impressed impulse, but instead, turning about, as by some command, retrace its course? Furthermore, how can this air, in so turning about, avoid being scattered into space, but instead impinge precisely on the notched end of the arrow and again push the arrow on and adhere to it? Such a view is quite incredible and borders rather on the fantastic. . . .

Now there is a second argument which holds that the air which is pushed in the first instance [i.e., when the arrow is first discharged] receives an impetus to motion, and moves with a more rapid motion than the natural [downward] motion of the missile, thus pushing the missile on while remaining always in contact with it until the motive force originally impressed on this portion of air is dissipated. This explanation, though apparently more plausible, is really no different from the first explanation by *antiperistasis,* and the following refutation will apply also to the explanation by *antiperistasis.*

In the first place we must address the following question to those who hold the views indicated: "When one projects a stone by force, is it by pushing the air behind the stone that one compels the latter to move in a direction contrary to its natural direction? Or does the thrower impart a motive force to the stone, too?" Now if he does not impart any such force to the stone, but moves the stone merely by pushing the air, and if the bowstring moves the arrow in the same way, of what advantage is it for the stone to be in contact with the hand, or for the bowstring to be in contact with the notched end of the arrow.

For it would be possible, without such contact, to place the arrow at the top of a stick, as it were on a thin line, and to place the stone in a similar way, and then, with countless

> machines, to set a large quantity of air in motion behind these bodies. Now it is evident that the greater the amount of air moved and the greater the force with which it is moved the more should this air push the arrow or stone, and the further should it hurl them. But the fact is that even if you place the arrow or stone upon a line or point quite devoid of thickness and set in motion all the air behind the projectile with all possible force, the projectile will not be moved the distance of a single cubit. (*Ibid.*, pp. 221-222.)

If the motion of the air does not move the projectile but rather resists the movement of the projectile, what is it that causes the continued motion of the projectile? Newtonian physics, rejecting the necessity of a continuing force, answers in terms of inertia. Philoponus' explanation is that an incorporeal kinetic force has been impressed in the body and this impressed force continues the movement of the body until it is spent by the resistance to movement present by the weight of the body and the resistance of the air.

> From these considerations and from many others we may see how impossible it is for forced motion to be caused in the way indicated. *Rather is it necessary to assume that some incorporeal motive force is imparted by the projector to the projectile,* and that the air set in motion contributes either nothing at all or else very little to this motion of the projectile. If, then, forced motion is produced as I have suggested, it is quite evident that if one imparts motion "contrary to nature" or forced motion to an arrow or a stone the same degree of motion will be produced much more readily in a void than in a plenum. And there will be no need of any agency external to the projector. . . . (*Ibid.*, p. 223.)

Like his criticism of Aristotle's views of natural movement, Philoponus' concept of an incorporeal kinetic force as the source of the continuing movement of a projectile had considerable historical influence, particularly among the philosophers of Islam, such as Avicenna, Alpetragius, and Abu 'l-Barakat, and possibly also indirectly among the Latin schoolmen of the late thirteenth and fourteenth centuries. We shall examine in a later volume the various emendations of Philoponus' theory of impressed force, which form an interesting background to the rise of a concept of momentum in early modern times. It is supposed that Philoponus did not originate the theory of an impressed force, that it went back to the astronomer Hipparchus in the second century B.C. or even further. But it should be pointed out that the earliest *extant* account of the theory is that of our sixth-century Neo-Platonist.

Scarcely less important and influential as a critic of Aristotle was Philoponus' contemporary Simplicius, whose direct influence on the Latin West began with translations made in the thirteenth century. We have already observed Simplicius' importance in transmitting some of the contents of Theophrastus' treatise on the opinions of the early natural philosophers. To Simplicius we owe also the longest extant passage describing the astronomical theory of concentric spheres developed by Eudoxus and adopted by Aristotle, as well as Strato's observations on the acceleration of falling bodies (see Chap. 6). Like Philoponus, Simplicius on more than one occasion appealed to the evidence of experience and experiment to oppose or confirm a scientific judgment. His attempt to settle the question of whether air or water has "weight in its own place" by an experiment is noteworthy although indecisive.

Now Ptolemy seeks to prove the proposition that air has no weight in its own medium by the same experiment of the inflated skin. Not only does he contradict Aristotle's view that the skin when inflated is heavier than when uninflated, but he maintains that the inflated skin is actually lighter.

I performed the experiment with the greatest possible care and found that the weight of the skin when inflated and uninflated was the same. One of my predecessors who tried the experiment wrote that he found the weights to be the same, or rather that the skin was a trifle heavier before inflation, a result which agrees with that of Ptolemy.

Now if the result of my experiment is correct, it follows,

clearly, that in their respective natural places the elements are without weight, having neither heaviness nor lightness. . . . (*Ibid.*, p. 248.)

One of the most influential sections of Simplicius' commentary on Aristotle's *Book of the Heavens,* translated in the thirteenth century, is his summary of the various attempts to explain the acceleration of falling bodies, and the theories that he describes have considerable popularity in the fourteenth century and later. The theories described by Simplicius are reducible to three main explanations.

[1] Acceleration takes place because a body nearing its natural place is desirous of completing or perfecting its form as fast as possible. "As bodies approach the whole mass of their own element they acquire a greater force therefrom and receive their own form more perfectly; that thus it is by reason of an increase of weight that earth moves more swiftly when it is near the center."

[2] Hipparchus, on the other hand, in his work entitled *On Bodies Carried Down by Their Weight* declares that in the case of earth thrown upward it is the projecting force that is the cause of the upward motion, so long as the projecting force overpowers the downward tendency of the projectile, and that to the extent that this projecting force predominates, the object moves more swiftly upwards; then, as this force is diminished (1) the upward motion proceeds but no longer at the same rate, (2) the body moves downward under the influence of its own internal impulse, even though the original projecting force lingers in some measure, and (3) as this force continues to diminish the object moves downward more swiftly, and most swiftly when this force is entirely lost.

Now Hipparchus asserts that the same cause operates in the case of bodies let fall from above. For, he says, the force which held them back remains with them up to a certain point, and this is the restraining factor which accounts for the slower movement at the start of the fall. . . . (*Ibid.*, p. 209.)

[3] Now there are not a few who assert that bodies move downward more swiftly as they draw nearer their goal because objects higher up are supported by a greater quantity of air, objects lower down by a lesser quantity, and that heavier objects fall more swiftly because they divide the underlying air more easily. For just as in the case of bodies which sink in water the lighter they are the more does the water seem to hold them up and resist the downward motion, so it is fair to suppose that the same thing happens in air, and that the greater the amount of underlying air, the more do lighter objects seem buoyed up. . . . (*Ibid.*, p. 210.)

It should be noted that the second of these theories, the theory of Hipparchus, was the one adopted by Galileo in his youth.

III

In our discussions of the Latin science we mentioned the popularity of the encyclopedia and the handbook. Together with the commentary, these forms of scientific literature also became popular in the Greek East. Thus the corpus of literature in a branch of science in the East consisted of certain selected "classical" Hellenistic or earlier works, commentaries, compendious encyclopedias, and handbooks or epitomes. The organization of medical literature in the sixth century was illustrative of this variety of literary forms. To the classical works of Hippocrates and Galen were added numerous compendia, epitomes, and occasionally an original commentary. Thorndike has noticed the trend in medicine toward compendia.

As Tribonian and Justinian boiled down the voluminous legal literature of Rome into one *Digest*, so there was a similar tendency to reduce the past medical writings of the Greeks into one compendious work. Paul of Aegina, writing in the seventh century, observes in his preface that it is not right, when lawyers who usually have plenty of time to reflect over their cases have handy summaries of their subject to which they can refer, that physicians whose cases often require immediate action should not also have some convenient handbook, and the more so since many of them are called upon to exercise their profession not in large cities with easy access to libraries, but in the country, in desert places, or on shipboard. Oribasius, friend and physician of the emperor Julian, A.D. 361-363, had made such a compendium by that emperor's order. In this he embodied so much of Galen's teachings that he became known as "the ape of Galen," although he also used more recent writers. But Paul of Aegina regarded this work of Oribasius as too bulky, since it originally comprised seventy-two books although only twenty-five are now extant, and so essayed a briefer compilation of his own. . . . (L. Thorndike, *A History of Magic and Experimental Science,* I, New York, 1923, pp. 568-569.[2])

Adoption of these less specialized forms does not necessarily mean sterility and lack of originality. Critical ability and originality are not entirely wanting in the great medical compilers of the age: Aetius of Amide (sixth century), Alexander of Tralles (sixth century), and Paul of Aegina (seventh century). Thus Alexander's principal work, his *Handbook,* has been described by one historian as "more a record of his own medical observations and experiences than a compilation from past writers." Without detailing the medical activity of these three and other post-classical physicians, we should observe that medical instruction as organized by and around these men at Alexandria and elsewhere had great influence on Islamic medicine. It is another case of medieval investigators' looking at classical material as *reorganized* and *selected* by the physicians or philosophers of late antiquity.

IV

One final event in the course of late antique Greek science needs cataloguing: the preliminary and partial translation of the Greek corpus of scientific learning into Syriac. Just as parts of the Greek corpus were turned into Latin, the common language of the Western half of the Roman Empire, so certain

[2] Quoted by permission of the publisher, the Columbia University Press.

Greek logical, medical, alchemical, and astronomical works began in late antiquity to be translated into Syriac, a Semitic tongue common to divers localities and peoples of the Eastern parts of the Empire. The composition of independent philosophical works in the Syriac language goes back at least to the second century. But the translating activity in which we are interested at this point of our studies arises in the Nestorian school of Edessa (northern Mesopotamia) in the fifth century. The rise of a "national" Syrian church or Christian Syrian sect with formal theological education in Syriac provided the initial stimulus to the translation of some of the logical works of Aristotle into Syriac. We cannot go into the religious doctrines of the Nestorian movement, which became firmly entrenched in certain Syrian cities by the fifth century; but we can insist that the conscious attempt to preserve the Nestorian doctrines did much to nourish and stimulate further a separate development of Syriac Hellenism.

Without discussing in any detail the fifth-century translators into Syriac, we can note that they knew and rendered into Syriac Aristotle's more elementary logical texts, particularly *The Categories* and the *On Interpretation*. It is also probable that the *Prior Analytics* was translated in the fifth century.

The increasing hostility toward the Nestorians of Edessa resulted in their final expulsion from that city in 489. At least some of those expelled went across the Persian border to the town of Nisibis (somewhat to the east of Edessa, in northern Mesopotamia). In the course of the sixth century, Syriac schools took root and flourished in several Persian towns. The Syriac school that was to have most influence on the development of Arabic learning was that at the great Persian center Jundi-Shapur (modern town of Shah Abad?). Jundi-Shapur had developed into a flourishing intellectual center of Hellenism by the sixth century. Two centuries later, after the Arabs conquered Syria and Persia and built their new capital of Baghdad, they called upon the Syriac Christians of Jundi-Shapur to provide them with the best medical and linguistic talent available in their empire. Jundi-Shapur had become by that time one of the most famous centers of medical teaching in the Near East. There seems little doubt that its curriculum was based on the medical teaching of Alexandria, and it is probably to such schools as that at Jundi-Shapur that the great Syriac and Arabic translator Hunain ibn Ishāq was referring when he alluded to "Christian" schools of his day.

> These are the books (of Galen) to the reading of which the students in the medical school at Alexandria were confined. . . . They were accustomed to meet every day for the reading and interpretation of one of the standard works, in the same way in which, in our days, our Christian friends are accustomed to meet every day at the educational institution known as σχολή for the study of a standard work from among the books of the Ancients. Concerning the remainder of (Galen's) books they were accustomed to read them everyone for himself, after an introductory study of the aforementioned books; just as our friends read to-day the explanations of the books of the Ancients. (*Isis*, Vol. 8, p. 702.)

It should be pointed out that not all of the Syriac activity grew out of the Nestorian schools. Some Syriac-speaking peoples took up the views of the Monophysites, a sect rather directly opposed to the Nestorian views on Christ's nature. In fact, the best of the early Syriac translators was not a Nestorian but a Monophysite. This was Sergius of Reshaina (d. *ca.* 536?). Sergius was a Christian priest and physician whose activity as a translator was marked. His home of Reshaina is located about halfway between Edessa and Nisibis, in northern Mesopotamia; and he was no doubt acquainted with the neighboring Nestorians, for he dedicates a work to one of their bishops. His translation of the pseudo-Aristotelian book *On the Universe* (*De mundo*) has been praised for its skilful rendering of the Greek. He also translated *The Categories* of Aristotle as well as the Neo-Platonic *Introduction to the Categories* of Porphyry. Thus he was translating this Aristotelian logical material into Syriac at almost the same time as Boethius was turning it into Latin at the other end of the Empire.

But Sergius' most important translations were of medical works. Although the extant copies of these medical translations are few, from the detailed remarks of the great Syriac-Arabic translator of the ninth century, Hunain ibn Ishāq, we learn the titles of some twenty-six works of Galen translated by Sergius. Sergius was particularly interested in translating the works of Galen which served as textbooks in the school of Alexandria, where Sergius himself had studied. Hunain ibn Ishāq is often quite critical of Sergius' translations. For example, of his translation of Galen on *The Kinds of Fevers,* Hunain says, "Sergius translated this book, but not in a creditable manner." Although he makes other such remarks about Sergius' incompetent translations, he occasionally speaks of his later translations in a more favorable light; thus, when speaking of one Galenic translation of Sergius, Hunain says: "Sergius translated this book into Syriac, the first parts when he was yet weak and inexperienced in translation work. He translated the remaining eight parts when he had acquired experience so he did this version better than that of the first six parts."

Whether or not Sergius' translations are bad, it must be remembered that he was probably the first significant translator of Greek medicine into a Semitic language. On his own he also composed two works, *On the Influence of the Moon* and *The Movement of the Sun.* They no doubt depended on Greek sources. Sergius has been epitomized by a later author as "a man eloquent and greatly skilled in the books of the Greeks and Syrians, and a most learned physician of men's bodies. He was orthodox in his opinions . . . but his morals corrupt, depraved and stained with lust and avarice."

Our consideration of the other Syriac translators prior to the Arabic period can be limited to one: the celebrated seventh-century translator Severus Sebokt. Sebokt was a bishop and monk about 640 at the convent of Kenesre on the Euphrates, a monastery that had become celebrated for its teaching of Greek. Like his predecessors he continued the study of Aristotelian logic. But more important for us, he pursued astronomical studies. He may have been the author of an extant Syriac translation of Ptolemy's *Almagest.* In addition, he composed a work on the phases of the moon; one on constellations, which is a rambling work including geography, chronology, and cosmology; a tract on the astrolabe (one of the chief astronomical instruments of the day); and so on. He makes mention of the Hindu numerals (i.e., the numerals known to us ordinarily as the Arabic numerals) in a work dated before 660. All in all, Severus appears to have been a key figure in the partial transmission of Greek learning into Syriac.

Before leaving our discussion of Syriac translations we should note that one of the largest sources of extant Syriac letters is a corpus of alchemical works, works which are either translations from the Greek or paraphrases of previous Greek alchemical works. We shall have occasion to mention these texts when we speak in a later volume of Islamic alchemy.

Chapter 14

In the End the Beginning

WE HAVE now reached the point where the Greek corpus of scientific learning had run its course in antiquity. We have seen that by the year 600 preliminary efforts had been made to turn parts of the corpus into Latin in the West and into Syriac in the East. But as yet not enough of the corpus had been translated to constitute a major conversion of Greek science into another language. We have also seen that in the East—in several centers of Hellenistic learning—most of the best classical scientific authors, such as Aristotle, Archimedes, Euclid, Apollonius, Hero, Ptolemy, Galen, continued to be studied and were in fact re-edited and commented upon as late as the fifth and sixth centuries.

Now the succeeding history of Greek science would show that it was converted first into the Arabic tongue, in the ninth and tenth centuries, and then once again from both Arabic and Greek into Latin, in the twelfth and thirteenth centuries. But we must set that history aside as the subject of another volume.

We shall do well to remember as we close our preliminary account of Greek science in antiquity that, although the Greek corpus received major additions and modifications at the hands of the Arabic and Latin authors, it was still essentially the Greek learning that made its way to the Latin West. By the thirteenth and fourteenth centuries it constituted the major element of scientific studies. For example, it was the interpenetration of Euclidian-Archimedean mathematics and Aristotelian philosophy in the schools that accounted for the major achievements of the medieval mechanicians in the thirteenth and fourteenth centuries.

Furthermore, it was still basically the Greek scientific corpus that continued to be studied in the West in the later Middle Ages and early modern times. And it was in great part that corpus which stimulated the most fertile scientific activity in the sixteenth and seventeenth centuries. Thus Vesalius started from Galen and Benedetti and Galileo from Archimedes and Apollonius.

It was Seneca who insisted that science builds upon past foundations. And surely the foundation stones quarried, faced, and laid in Greek antiquity have provided the surest support for the scientific structure raised in modern times.

Appendix I

Archimedes and the Quadrature of the Circle

(Proposition I from Archimedes' *Measurement of the Circle*—A Free Paraphrase)

Introduction: In the earliest period of mathematics, formulae for the areas of figures bound by straight lines (rectilinear figures) were worked out; e.g., for a triangle, $\frac{1}{2}a \cdot b$, or a rectangle $a \cdot b$, a and b being straight lines, of course. Now when the problem arose of finding a formula for the area of figures bound by curved lines (curvilinear figures), the whole question of "quadrature" was raised. "Quadrature" is simply the finding of a rectilinear figure equivalent to the curvilinear figure. Archimedes shows in the fol-

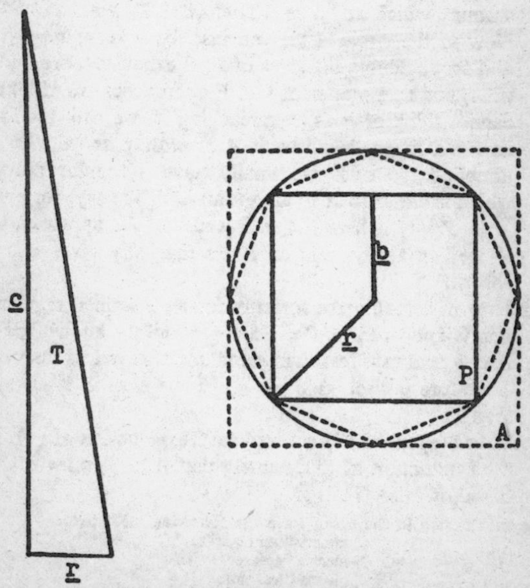

FIG. 28. The quadrature of the circle.

lowing proof that a right triangle, of which the sides including the right angle are equal respectively to the circumference and radius of a circle, is equal in area to the circle. This triangle is the rectilinear figure equal to the curvilinear figure (the circle) the area of which is sought.

Proof: Suppose that A = area of a circle.
T = area of a right triangle with sides c and r, the circumference and radius of the circle.
P = area of an inscribed regular polygon, of which the perimeter is s and the line from the center of the circle or figure to the middle of one of its sides is b.

To prove: $A = T$.*

1. Either $A = T$ or $A \neq T$.
2. If $A = T$, we have the proposition; if $A \neq T$, then $A > T$ or $A < T$.
3. Suppose first that $A > T$. Then $A > T$ by some definite amount, which we let be e. Then $A = T + e$.
4. Now in the circle A we can inscribe a regular polygon P, and as we double the sides of P, P exhausts more and more of A, and as a matter of fact P approaches A. Although we cannot think of P as ever reaching A, we can, by doubling its sides, make P approach A *as closely as we like.* So by doubling the sides we finally have a regular polygon P which is closer to A in area than is T. We say, for example, $A = P + f$, where $f < e$. Because of P's approach toward A we can always find an f less than any e we assume. In short, $P > T$.
5. Now in actuality the formula for an inscribed regular polygon is known to be $P = \frac{1}{2}\, b \cdot s$; and we know $T = \frac{1}{2}\, cr$.
6. And b is always less than r and s is always less than c.
7. Therefore in fact $\frac{1}{2}\, b \cdot s < \frac{1}{2}\, c \cdot r$, or P is always *less* than T.
8. Thus the fact of (7) contradicts the deduction of (4). Hence the assumption of (3), namely that $A > T$, on which (4) is based, is false. Thus $A \ngtr T$.

* The symbols used in the proof have the following meanings:
$\neq$ means "not equal to."
$>$ means "greater than."
$<$ means "less than."
$\ngtr$ means "not greater than."
$\nless$ means "not less than."

9. In a series of similar steps we can show by the use of a circumscribed polygon that the assumption $A < T$ is also false. Thus $A \nless T$.
10. Thus, if $A \ngtr T$ and $A \nless T$, then $A = T$. Q.E.D.

Appendix II

Archimedes and the Application of Mechanics to Geometry

(Proposition I from Archimedes' *Method—A Paraphrase*)

1. *Enunciation* (*protasis*): Any segment of a right-angled cone (i.e., of a parabola) is 4⁄3 of the triangle which has the same base and equal height.
2. *Example* (*ekthesis*): Let the segment be *APBC* and the triangle *ABC*.
3. *The specification of the proposition* (*diorismos*): Segment *APBC* = 4⁄3 triangle *ABC*.

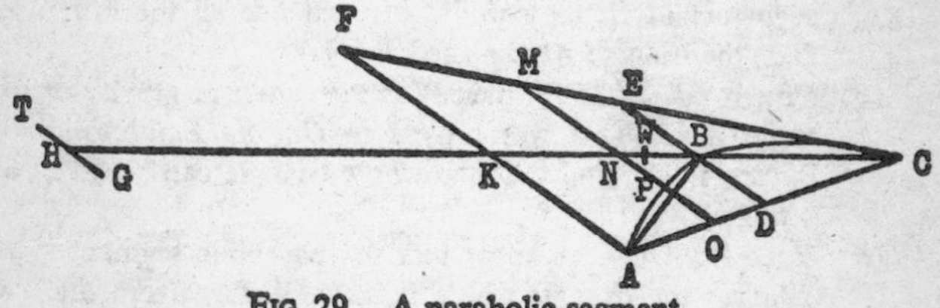

Fig. 29. A parabolic segment.

4. *Further construction* (*kataskeue*):
   (*a*) *P* is any point on the parabola *APBC;* let *BD* be drawn parallel to the axis.
   (*b*) Extend a tangent from *C* indefinitely.
   (*c*) Extend diameter *BD* to tangent forming line *DBE*.
   (*d*) Draw line *MNPO* through point *P* and parallel to *DBE*.
   (*e*) Draw line *AKF* through *A* and parallel to *DBE*.
   (*f*) Extend *BC* to cut *MO* at *N*, *FA* at *K*, and further so that *HK* = *KC*.
   (*g*) Then let *HC* be considered a balance beam with fulcrum at *K*.

5. *Proof* (*apodeiksis*):
   (*a*) Since *APBC* is a parabola and *EC* lies on the tangent, and *DC* is a semiordinate, then $BD = EB$. [This was proved in the *Elements of Conics* by Euclid.]
   (*b*) Then $MN = NO$,
   $FK = KA$ [Euclid, VI, 4, and V, 9—*MO* and *FA* being parallel to *ED* and using (*a*)].
   (*c*) And $AC : AO = MO : PO$ [Property of parabola, see Quad. parab. 5 and Euclid VI, 18].
   (*d*) Thus $MO : PO = KC : KN$ (since $AC : AD = KC : KN$ by similar triangles).
   $= HK : KN$ (since $KC = HK$ by construction).
   (*e*) Then if we consider *HC* as a balance and leave line *MO* suspended at its midpoint *N*, which is its center of gravity, and if we take a segment $TG = PO$ and suspend it at *H*, then from (*d*), and the law of the lever, *TG* at *H* just balances *MO* at *N*.
   (*f*) But *P* is any point and *PO* is any line in the segment, while *MO* is any corresponding line in the triangle *AFC*. Thus there are as many lines in the segment as in the triangle. Hence if we place all lines *PO* (i.e., the whole parabolic segment) at *H*, they would just balance all the lines *MO* (i.e., the triangle *AFC* placed as it is).
   (*g*) Now if $CK = 3KW$, then *W* is the center of gravity of triangle *AFC* [This was proved in *On the Equilibrium of Planes*, I, 15], and thus the whole triangle can be taken as acting at *W*.
   (*h*) Since from (*f*) we know that the parabolic segment at *H* balances the triangle where it is, we deduce from the law of the lever that
   segment $APBC . HK =$ triangle $AFC . KW$.
   But $HK = CK = 3KW$, and so
   segment $APBC = \frac{1}{3}$ triangle *AFC*.
   And it can be shown that triangle $AFC = 4$.triangle *ABC*.
   (*i*) And so segment $APBC = \frac{4}{3}$ triangle *ABC*.
6. *Conclusion* (*symperasma*): And so a parabolic segment is $\frac{4}{3}$ of the triangle with the same base and altitude.

The reader is reminded of three important points in connection with this proof: (1) Archimedes applies a mechanical law, the law of the lever, for the determination of a geometric theorem, or perhaps it would be better to say for the "discovery" of a geometric theorem. (2) And in order to apply mechanical reasoning to geometry, Archimedes assumes further that the center of magnitude of a figure can be taken as its center of gravity, so that the whole magnitude can be said to be concentrated, so far as weight is concerned, at the center of magnitude. (3) A magnitude can be conceived as being composed of an infinite number of line segments; or, perhaps better, any two area magnitudes with the same base can be thought of as being composed of the same number of segments.

We have already noted in Chapter 5 the statement by Archimedes that he did not consider this method demonstrative, but only heuristic.

Note: The formal Greek terms for the parts of a "Euclidian" proof have been added as illustration. They are not in the Greek text.

Appendix III

Apollonius of Perga and the Development of Conic Sections

(Cf. T. L. Heath, *History of Greek Mathematics*, II, Oxford, 1921, pp. 110-141)

Certainly one of the earliest authors to use conic sections in geometric problems was the pupil of Eudoxus, Menaechmus, who flourished about 360-350 B.C. It is believed by Heath that Menaechmus generated the three conic sections from a right circular cone the vertical angle of which was a right angle. Thus a parabola can be produced by passing a plane at right angles to one of the generators of the cone. A hyperbola results when a plane parallel to the axis of the cone cuts the cone. And, finally, an ellipse can be formed by a plane which cuts all the generators of the cone, but is not parallel to the base. But regardless of the way in which Menaechmus formed his conic sections, a conventional way had grown up by the time of Euclid and Archimedes. For each section a right circular cone was used and the cutting plane remained perpendicular to one of the generators, *but the vertical angles were varied.* With the vertical angle a right angle a parabola was produced, with an acute angle an ellipse was produced, and with an obtuse angle a hyperbola was formed. Hence a parabola was called "a section of a right-angled cone," an ellipse a "section of an acute-angled cone," and a hyperbola a "section of an obtuse-angled cone."

It is known that Menaechmus had used the properties of conic sections to solve the problem of finding two mean proportional lines between two given lines, a problem which we rewrite in modern fashion,

(1) $\frac{a}{x} = \frac{x}{y} = \frac{y}{b}$,

where a and b are the given quantities, and x and y are the unknowns. We can rewrite (1) as follows:

(2) $x^2 = ay$, $y^2 = bx$, $xy = ab$.

It is obvious that these are three simultaneous equations and that the first two are the equations of parabolas and the third is an equation of a hyperbola, using a modern rectangular coordinate

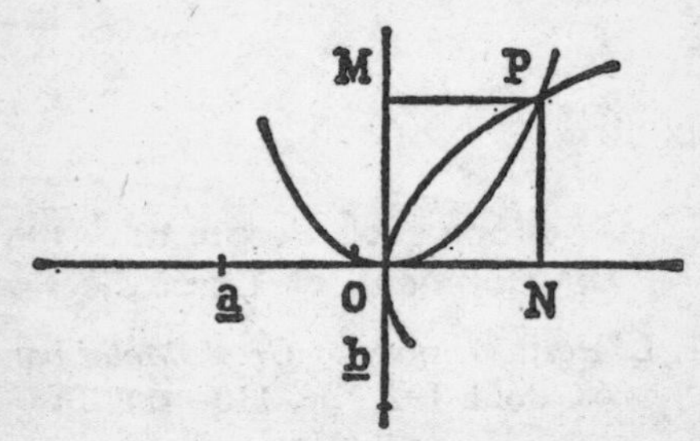

FIG. 30. Intersecting parabolas.

system. Hence it is equally obvious that we can solve for x and y by using any *two* of the equations—or, to put it another way, that the two mean proportionals can be found by the intersection of the two parabolas, or by either of the parabolas with the hyperbola. It was precisely by such intersections that Menaechmus solved the problem. For supposing he uses the line a as a parameter to construct one parabola and the line b as a parameter to construct the other parabola, the parabolas being respectively about the y and x axes, and O being the common vertex. Then, as shown in the accompanying figure, P is the point of intersection of the two parabolas, and PM and PN are the desired values of x and y.

Among the authors after Menaechmus who contributed to the development of conic sections we can single out Aristaeus (flourished somewhat earlier than Euclid), Euclid, who composed a treatise in four books on conics, and Archimedes. But all of the prior work was superseded by the *Conics* of Apollonius of Perga, some twenty-five years the junior of Archimedes. It was largely from this work of Apollonius that early modern scientists learned their geometry of conic sections. We cannot in this brief appendix discuss in detail the many theorems of this work (of which Books I-IV only are extant in Greek, V-VII in Arabic, and VIII is lost). It will suffice to show how he expresses the fundamental properties

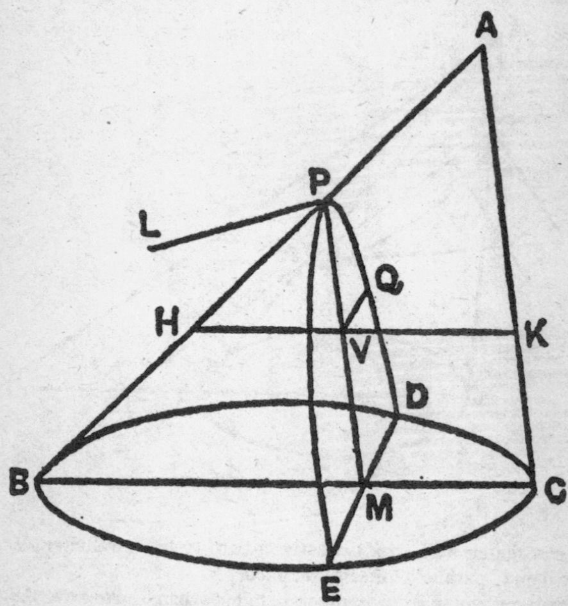

FIG. 31. The parabola.

of the different sections in a new way, so that he arrives at the new terms "parabola," "hyperbola," and "ellipse." For example, let us take the parabola. Referring to Fig. 31, we can note that line *PL* is the parameter of the parabola — i.e., the constant quantity that expresses the nature of the parabola. It is constructed as a line perpendicular to the axis of the parabola.

He then assumes for *PL* (1) $PL : PA = BC^2 : (BA \cdot AC)$. Then *HK* is made parallel to *BC*, and thus is a diameter of circle *HQK*. Hence, (2) $QV^2 = HV \cdot VK$. Then (3) $HV : VP = BC :$

CA, and (4) $VK : PA = BC : BA$. Multiplying (3) by (4), we have (5) $(HV \cdot VK) : (VP \cdot PA) = BC^2 : (CA \cdot BA)$. Then substituting (1) and (2) in (5), we have (6) $QV^2 = PL \cdot PV$. The equation (6) represents the "equation" of the parabola. If we consider PV the abscissa (x) and QV the ordinate (y), and let the parameter PL be called p, then we have the familiar form: $y^2 = px$. Now for Apollonius PL is considered the "line of application," and thus for a parabola the square of the ordinate (QV^2) is equal

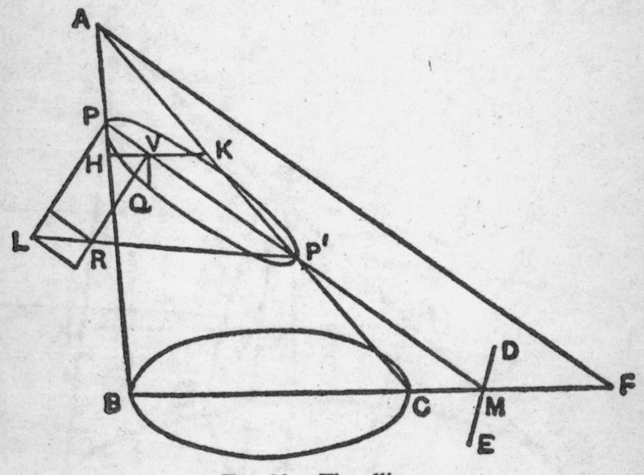

FIG. 32. The ellipse.

to a rectangle ($PL \cdot PV$) exactly *applied* to the parameter PL, and the name "parabola" means "applied."

Without going into as much detail we can show how the new application method of expressing the fundamental properties applies also to the ellipse and the hyperbola. For an ellipse, Apollonius assumes a parameter PL defined as follows: $PL : PP' = (BF \cdot FC) : AF^2$ (see Fig. 32). From this he shows by a series of steps that $QV^2 = PV \cdot VR$. And so in the case of the ellipse the square of the ordinate (QV^2) is equal to a rectangle ($PV \cdot VR$) applied to PL but which *falls short* of the rectangle $PL \cdot PV$ by the amount of a small parallelogram LR which is similar to parallelogram $LP \cdot PP'$ and is similarly situated. And the fact that we have a rectangle which *falls short* of the exact application to PL determines the name "ellipse," which means "falling short."

Finally, assuming that $PL : PP' = (BF \cdot FC) : AF^2$ (see Fig. 33), Apollonius shows that, for a hyperbola, $QV^2 = PV \cdot VR$. Thus in the case of a hyperbola the square of the ordinate (QV^2) is equal to a rectangle $PV \cdot VR$ which *exceeds* the rectangle $PL \cdot PV$ by the small rectangle LR similar to $LP \cdot PP'$ and similarly situated. From the fact that $PV \cdot VR$ *exceeds* the exact application

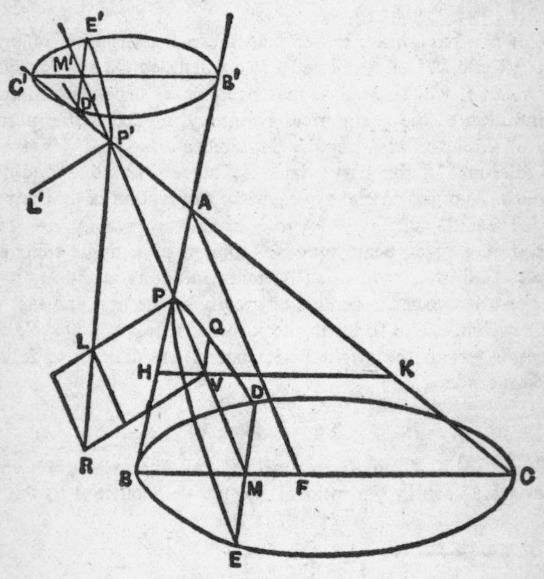

FIG. 33. The hyperbola.

$PL \cdot PV$, it is called a hyperbola, which means "exceeding." And so all of the conic sections are expressed as equations of areas in which one side of the equation is the square of the ordinate and the other is a rectangle applied equally, or in deficiency, or in excess to a given line, the parameter, called also the *latus rectum*. Hence by using the method of application of areas, which we saw in Chapter 5 went back to the Pythagoreans, Apollonius has managed to generalize completely the properties of conic sections.

Appendix IV

Archimedes' Proof of the Law of the Lever

This is a paraphrase rather than a direct translation of propositions VI and VII of Archimedes' work entitled *On the Equilibrium of Planes*. It will be noticed that Archimedes depends primarily on an intuition of the geometrical symmetry implied in the particular case of the lever where equal magnitudes are equally distant from the fulcrum. In the particular case he assumes that equilibrium pertains. Another crucial symmetrical assumption is that any number of equal magnitudes where centers of gravity are equally spaced on a given beam may be replaced by a single weight composed of all of the individual weights and hung at the midpoint of the beam between the centers of gravity of the first and last of the equal weights. The following exposition of this proof replaces the Greek letters of the original with letters less difficult to follow by modern readers.

Proposition VI

To prove, if *A* and *B* are unequal, but commensurable, magnitudes which are in the ratio of the linear distance *a* to the linear

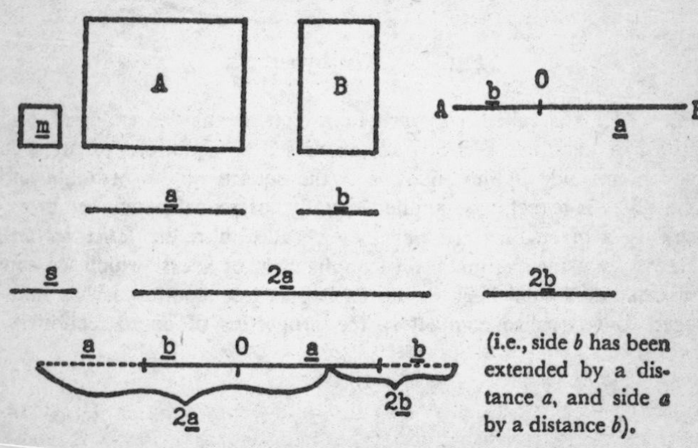

Fig. 34. Commensurable magnitudes on a lever.

distance b, and which are hung respectively on beam $a + b$ at distances from O respectively of b and a, that O is the center of gravity of the system and that the system therefore is in equilibrium.

(1) $\frac{A}{B} = \frac{a}{b} = \frac{2a}{2b}$ (given).

(2) Extend beam $a + b$ to form beam $2a + 2b$ as shown in the drawing.

(3) If A and B are commensurable, then a and b, as well as $2a$ and $2b$, are commensurable.

(4) Hence there is common measure of distance s and some common magnitude measurement of m, such that $k \cdot s = 2a$ and $k \cdot m = A$, and also such that $n \cdot s = 2b$ and $n \cdot m = B$. [From (1) and (3).]

(5) Now in the extended beam of (2) imagine that the n number of m's of B are hung at the midpoints of the n number of s's of $2b$. Similarly, take the k number of m's of A and distribute them at the midpoints of the k number of s's in $2a$. Thus we have distributed A and B uniformly over the beam $2a + 2b$.

(6) Since O is the center of beam $2a + 2b$ and since A and B are distributed uniformly over $2a + 2b$, O is the center of gravity of the magnitudes so distributed. But the magnitudes A and B so distributed act the same as if they were hung at the ends of b and a. Hence O is also the center of gravity of the latter system, and thus the system is in equilibrium. Q.E.D.

PROPOSITION VII

To prove, if $A + e$ and B are incommensurable magnitudes related as distance a to distance b and are hung at distances b and a respectively from O, that O is the center of gravity and thus the system is in equilibrium.

(1) $\frac{A + e}{B} = \frac{a}{b}$ (given).

(2) Either $A + e$ balances B, or it does not. If it does not, then either $A + e$ is depressed or it is elevated.

(3) Suppose that $A + e$ is depressed.

(4) Take away from $A + e$ a quantity e, *not* large enough to prevent the depression of A but such as to leave A commensurable with B. [*Note*: This is an application of the fer-

tile method of exhaustion used by Archimedes in proving geometric theorems (see Appendix I.)]

(5) Since we assumed that $\frac{A + e}{B} = \frac{a}{b}$, then $\frac{A}{B} < \frac{a}{b}$.

(6) From proposition VI when $\frac{A}{B} = \frac{a}{b}$, then equilibrium exists.

But in (5) it is said that $\frac{A}{B} < \frac{a}{b}$. Hence A is not sufficient to balance B, and so A will be *elevated.*

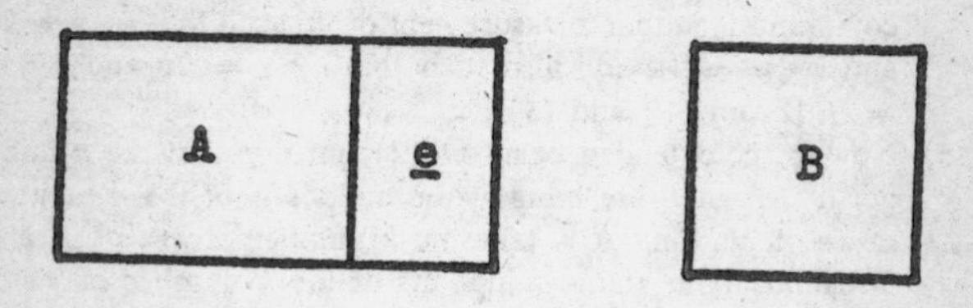

FIG. 35. Incommensurable magnitudes on a lever.

(7) But the fact of (6) contradicts the supposition of (4) and thus of (3), and hence (3)—namely, that $A + e$ is depressed—is false.

(8) In the same way it can be shown that $A + e$ is not elevated; and thus equilibrium exists. Q.E.D.

Appendix V

Ptolemy's Table of Chord Lengths

(See the *Mathematical Syntaxis,* Book I, Chap. IX)

Although it is conventional to begin the history of Greek trigonometry with a reference to the lost work of Hipparchus *On the Chords in a Circle,* trigonometry quite obviously had a prehistory in the development of astronomy, and particularly in the development of spherical geometry as applied to astronomy. It is evident also, as we have pointed out in Chapter 7, that Aristarchus made approximations of line lengths that were essentially measures of angles and thus were trigonometric in character. But assuming that trigonometry proper is not established until we have a *table* relating line lengths to angles, then we do well to follow the traditional approach and start with Hipparchus. For without doubt Hipparchus' lost work contained a table of chord lengths to be used for the solution of plane and spherical triangles. It is known, furthermore, that in the first century Menelaus (see Chap. 9), one of the best of the spherical geometers, composed a table of chord lengths. Thus Ptolemy in the next century was the heir of this already established tradition when he set out in the first book of the *Mathematical Syntaxis* to obtain a table of chord lengths. He found such lengths for every angle from ½° to 180°, proceeding by steps of ½°; these lengths he expressed in terms of parts of the diameter, which was assumed as having 120 parts.

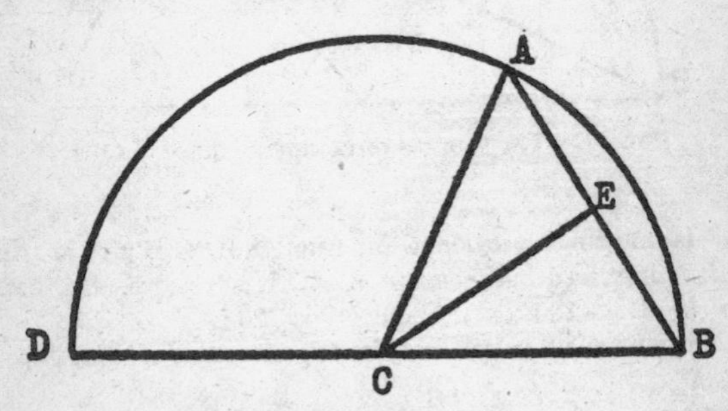

FIG. 36. Chords and sines.

The modern student will note that this table is equivalent to a table of the sines of one half of the angles. For by reference to the accompanying diagram one sees that the chord AB is the same number of parts of the diameter DB as EB is of CB. But AB expressed as parts of DB is the chord of a and EB expressed as parts of CB is the sine of $\frac{a}{2}$. This identity follows, since $DB = 2CB$ and $AB = 2EB$.

In setting out to build up a table of chords, Ptolemy sought first to find a few readily determined chords and to establish a few key geometrical relationships between chords of related angles.

(1) First by simple geometry and the calculation of certain square-root approximations he found the following chord lengths which are the sides of regular inscribed polygons:*

crd 36° = 37p 4′ 55″
crd 72° = 70p 32′ 3″
crd 60° = 60p — —
crd 90° = 84p 51′ 10″
crd 120° = 103p 55′ 23″

(2) Ptolemy next established by the Pythagorean theorem the relationship:

$$\text{crd}^2\, a + \text{crd}^2\, (180° - a) = 120^2$$

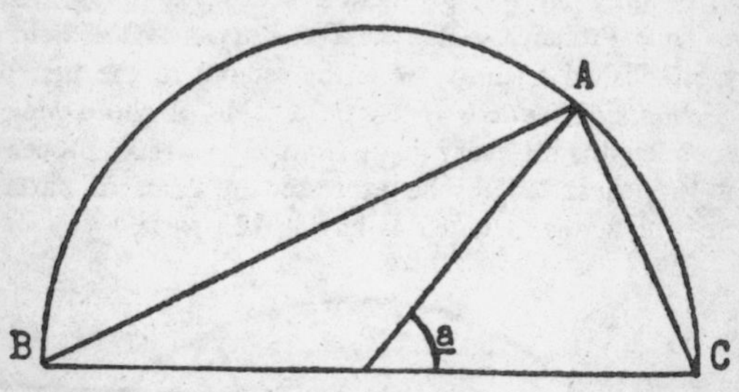

FIG. 37. The squares of complementary chords.

This is immediately evident, for triangle *ABC* is a right triangle since it is inscribed in a semicircle, *AC* is the chord of *a*, and *AB* is the chord of (180° − *a*), and *BC* is 120.

This is equivalent to the familiar trigonometric relationship

$\text{Sin}^2\, c + \text{Cos}^2\, c = 1$,

where $a = 2c$.

Now from this relationship Ptolemy observed that one could compute from the chord of *a* the chord of (180° − *a*), and he computed, for example,

crd (180° − 36°) = crd 144° = 114p 7′ 37″.

(3) After proving by geometry that the product of the diagonals of an inscribed quadrilateral equals the sum of the products of the opposite sides taken two at a time ("Ptolemy's theorem"), he established the following important relationship for the chord of the difference of two angles:

* p stands for one part of the diameter which contains 120 parts. The fractions are expressed as sexagesimals.

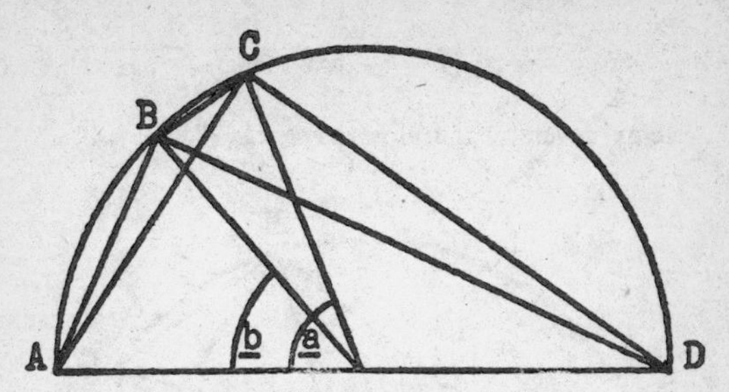

FIG. 38. The difference of chords.

$$\text{crd}\,(a-b) = \frac{\text{crd}\,a \cdot \text{crd}\,(180° - b) - \text{crd}\,b \cdot \text{crd}\,(180° - a).}{120}$$

This relationship will be evident from the following. In Figure 38 the following equivalent quantities should be noted:

$AB = \text{crd}\ b$ $\qquad BD = \text{crd}\ (180° - b)$
$AC = \text{crd}\ a$ $\qquad CD = \text{crd}\ (180° - a)$
$BC = \text{crd}\ (a - b)$ $\qquad AD = 120$

From "Ptolemy's theorem" concerning the diagonals of inscribed quadrilaterals, we get immediately the relationship:

$$AC \cdot BD = (BC \cdot AD) + (AB \cdot CD)$$

and thus

$$BC = \frac{(AC \cdot BD) - (AB \cdot CD)}{AD}$$

or

$$\text{crd}\,(a-b) = \frac{\text{crd}\,a \cdot \text{crd}\,(180° - b) - \text{crd}\,b \cdot \text{crd}\,(180° - a).}{120}$$

This is equivalent to the familiar formula

$$\sin\,(c - d) = (\sin c \cdot \cos d) - (\sin d \cdot \cos c)$$

where $\dfrac{a}{2} = c$ and $\dfrac{b}{2} = d.$

With this relationship we can find, e.g.,

$$\text{crd}\,(72° - 60°) = \text{crd}\ 12° = 12^p\ 32'\ 36''.$$

(4) The next important relationship developed by Ptolemy is the one for finding the chord of half of the angle when the chord of the angle is given. This he finds to be

$$\text{crd}\ \frac{a}{2} \quad \sqrt{60\ [120 - \text{crd}\ (180° - a)]}$$

which Ptolemy develops as follows (consult Fig. 39).

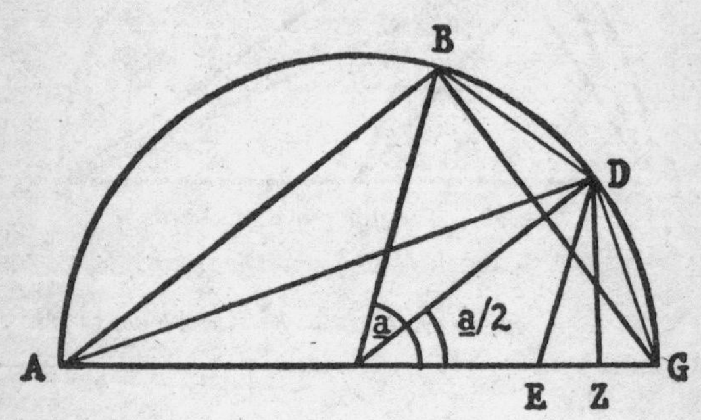

FIG. 39. The chord of half the angle.

Let us first note these equivalents:

$$GD = \text{crd}\ \frac{a}{2}$$
$$BG = \text{crd}\ a$$
$$AB = \text{crd}\ (180° - a)$$
$$AG = 120$$

Now GD is sought. BG is given and hence AB is given. AE is made equal to AB with DZ perpendicular to AG. Hence $ZG = ½ (AG - AB)$, for (1) $\triangle ABD = \triangle AED$, two sides and the included angle of one being equal to two sides and the included angle of the other; and thus (2) $DE = DB$; but as chords of equal angles $BD = DG$; and thus with $\triangle EDG$ isosceles and DZ a perpendicular to EG, (3) $EZ = ZG$ or $ZG = ½ (AG - AB)$.

But $AG : GD :: GD : GZ$ by similar triangles

or $GD^2 = AG \cdot GZ$

$= ½\ AG\ (AG - AB)$ where all the quantities are known.

And thus, substituting equivalent values,

$$GD = \sqrt{60\ [120 - \text{crd}\ (180° - a)]}.$$

The formula for GD^2 is equivalent to the well-known formulation

$$\sin^2 ½\ c = ½\ (1 - \cos c)$$

where $$c = \frac{a}{2}.$$

With this formulation, Ptolemy points out, many chords can be found as halves of known chords; e.g., the chord of 6° from the known chord of 12°, then from the chord of 6° the chord of 3°, and from it the chord of 1½°, and from it the chord of ¾°.

(5) Finally, Ptolemy gives us the most important rule of all, that for finding the chord of the sum of two angles, when the chord of each of the two angles is known. The proof is not unlike that for the determination of the chord of the difference of two angles, and makes use of "Ptolemy's theorem." Because of its similarity we give only the principal formulation, which is for the chord of the complementary angle of crd $(a + b)$, namely $(180° - a - b)$:

$$120 \cdot \text{crd}\,(180° - a - b) = \text{crd}\,(180° - a) \cdot \text{crd}\,(180° - b) - \text{crd}\,a \cdot \text{crd}\,b$$

[which formulation is equivalent to the familiar

$$\cos\,(c + d) = \cos c \cdot \cos d - \sin c \cdot \sin d$$

where $$c = \frac{a}{2} \text{ and } d = \frac{b}{2}$$]

It should be noted finally that from crd $(180° - a - b)$ we can immediately find the chord of its complementary angle, crd $(a + b)$, from the relationship given in (1). And this was sought.

(6) Following the formulation of (5) Ptolemy determined the approximation of the crd 1°, and then with (5) and this approximation the table was completely built up. For he found crd ½° using formula (4) on his approximation of 1°. Thus we simply have to determine all the succeeding angles by the addition of formula (5); e.g.:

crd ½° = op 31′ 25″
crd 1° = 1p 2′ 50″
crd 1½° = crd (1° + ½°) = 1p 34′ 15″
crd 2° = crd (3/2° + ½°) = 2p 5′ 40″
crd 2½° = crd (2° + ½°) = 2p 37′ 4″
etc.
crd 179° = crd (178½° + ½°) = 119p 59′ 41″
crd 179½° = crd (179° + ½°) = 119p 59′ 56″
crd 180° = 120 parts

As I have indicated, the principal use of the table of chords is in the calculation of spherical triangles. It is used, for example, in the problem of finding an arc between the equator and the ecliptic (Book I, Chap. XII). The reader is referred to M. Cohen and I. E. Drabkin, *A Source Book in Greek Science*, New York, 1948, pp. 84-85, for an example translated into English of the use of chords in the problem of determining the latitude of a place when there is known the length of the longest day of the year at that place.

Further Readings

The best general accounts covering either all or a large part of antique science are the following:

1. P. Brunet and A. Mieli, *Histoire des sciences: Antiquité*, Paris, 1935. (Includes selected readings.)
2. George Sarton, *A History of Science: (I) Ancient Science through the Golden Age*, Cambridge, Mass., 1952; *(II) Hellenistic Science and Culture in the Last Three Centuries B.C.*, Cambridge, 1959.
3. George Sarton, *Introduction to the History of Science*, I, Baltimore, 1927. (Contains extensive bibliographies and short biographies.)

On a more popular level are:

4. B. Farrington, *Science in Antiquity*, Oxford, 1936.
5. B. Farrington, *Greek Science: Its Meaning for Us*, 2 vols., London, 1944-1949. (Excellently written, but has a strong materialist bias.)
6. E. J. Dijksterhuis, *The Mechanization of the World View*, Oxford, 1961 (pp. 6-95).
7. J. L. Heiberg, *Mathematics and Physical Science in Classical Antiquity*, translated from the German by D. C. Macgregor, London, 1922.
8. W. A. Heidel, *The Heroic Age of Science* etc., Baltimore, 1933.
9. A. Reymond, *A History of the Sciences in Greco-Roman Antiquity*, translated from the French by R. G. de Bray, New York, 1927.
10. G. de Santillana, *The Origins of Scientific Thought*, Chicago, 1961.
11. S. Sambursky, *The Physical World of the Greeks*, New York, 1956 (Collier Books BS 28V).

The most useful single volume on Greek science is the following book of source readings: M. Cohen and I. E. Drabkin, *A Source Book in Greek Science*, New York, 1948. Those teachers who wish to use the present volume as a textbook will find that the Cohen-Drabkin volume will provide an excellent supplement. And the general reader will also find it profitable to have that volume at hand as he reads.

In suggesting further readings on the topics of each chapter I make no effort toward completeness. In fact, for the most part, I have merely added items of importance which are not found in the excellent bibliographies appended to Cohen and Drabkin's volume.

PART I

CHAPTER 1

Since Cohen and Drabkin have no bibliography for science in pre-history and in Egypt and Mesopotamia—the topic of this chapter—a somewhat longer list is given here than for the succeeding chapters:

1. J. H. Breasted, *The Edwin Smith Papyrus,* 2 vols., Chicago, 1930. (Contains excellent introductions evaluating Egyptian medicine.)
2. J. H. Breasted, *Dawn of Conscience,* New York, 1933.
3. A. B. Chace, *The Rhind Mathematical Papyrus,* 2 vols., Oberlin, Ohio, 1927-1929.
4. V. Gordon Childe, *Man Makes Himself,* London, 1936. (The best of Childe's numerous popularizations; has a materialist bias.)
5. G. Conteneau, *La Medicine en Assyrie et en Babylonie,* Paris, 1938. (Excellently illustrated; much more than a history of medicine.)
6. F. X. Kugler, *Sternkunde und Sterndienst in Babel,* 2 vols., and 3 suppl., 1907-1935. (The masterwork which recovered Babylonian astronomy.)
7. O. Neugebauer, *The Exact Sciences in Antiquity,* Princeton, 1951 (2nd edit., Providence, 1957). (By far the best single volume on Egyptian and Mesopotamian astronomy and mathematics; it includes excellent bibliographical supplements to its various chapters.)
8. O. Neugebauer and A. J. Sachs, *Mathematical Cuneiform Texts,* New Haven, 1945.
9. O. Neugebauer, *Astronomical Cuneiform Texts,* 3 vols., London, 1955.
10. R. A. Parker, *The Calendars of Ancient Egypt,* Chicago, 1950.
11. J. L. Partington, *The Origins and Development of Applied Chemistry,* London, 1935. (A detailed study of many aspects of ancient techniques; a superb summary of the modern literature on this subject.)
12. W. F. Petrie, *Wisdom of the Egyptians,* London, 1940. (The

mathematical sections can be used only with great caution, but this volume contains much technological information.)

13. H. Sigerist, *A History of Medicine*, I, New Haven, 1951. (The most recent general treatment based on the best available monographs.)
14. R. C. Thompson, *Dictionary of Assyrian Chemistry and Geology*, Oxford, 1936.
15. F. Thureau-Dangin, *Textes mathématiques babyloniens transcrits et traduits*, Leiden, 1938.

Chapters 2 and 3

Consult the bibliography in Cohen-Drabkin, pp. 560-561, and the following additional items:

1. F. M. Cornford, *Principium Sapientiae, The Origins of Greek Philosophical Thought*, Cambridge, Eng., 1952. (A radical departure from the Burnet "positivistic" treatment of the pre-Socratics.)
2. H. Diels and W. Kranz, *Die Fragmente der Vorsokratiker*, 9th edit., 3 vols., Berlin, 1952 (reprinted, 1960).
3. K. Freeman, *The Pre-Socratic Philosophers*, Oxford, 1946.
4. K. Freeman, *Ancilla to the Pre-Socratic Philosophers*, Oxford, 1948. (These works render much of Diels into English.)
5. W. Jaeger, *The Theology of the Early Greek Philosophers*, Oxford, 1947. (Also disagrees with the Burnet presentation of the pre-Socratics.)
6. G. S, Kirk and J. E. Raven, *The Presocratic Philosophers*, Cambridge, 1957.
7. G. Vlastos, "Theology and Philosophy in Early Greek Thought," *The Philosophical Quarterly*, Vol. 2 (1952), pp. 97-123. See also other brilliant articles by Vlastos in *Classical Philology*, 42 (1947), pp. 156-58; *The Philosophical Review*, Nov. & Jan., 1945; *ibidem* Jan., 1950; *American Journal of Philology*, 76 (1955), pp. 310-13, 337-68.

Chapter 4

Consult the bibliographies for biology, botany, zoology, and medicine in Cohen-Drabkin, pp. 566-568. To these bibliographies, add the following:

1. W. Heidel, *Hippocratic Medicine*, New York, 1941. (Emphasizes the relations of medicine with science; of uneven quality, but parts are worth reading.)
2. W. Jaeger, *Aristotle* etc., Oxford, 1934.
3. W. D. Ross, *Aristotle*, London, 1930.

Chapter 5

See the bibliography for mathematics in Cohen-Drabkin, pp. 561-563. Also useful is the posthumous volume of T. L. Heath, *Mathematics in Aristotle,* Oxford, 1949. A most interesting approach to the dependence of Greek mathematics on Babylonian linear methods is to be found in the sixth chapter of Neugebauer's *The Exact Sciences in Antiquity,* already cited in the readings for Chapter 1. Also of interest are the recent volumes: Paul-Henri Michel, *De Pythagore à Euclide,* Paris, 1950 and B. L. van der Waerden, *Science Awakening,* Groningen, 1954 (2nd edit., 1961).

Chapter 6

See the bibliography for Greek physics and technology in Cohen-Drabkin, pp. 564-565. Also consult:

1. Aristotle, *Physica* (in the edition of W. D. Ross, Oxford, 1936, which has rich and informative commentaries).
2. H. Canteron, *La Notion de force dans le système d'Aristote,* Paris, 1923.
3. F. L. Cooper, *Aristotle, Galileo, and the Tower of Pisa,* Ithaca, 1935.
4. A. G. Drachmann, *Ktesibios, Philon and Heron. A study in Ancient Pneumatics,* Copenhagen, 1948. (A new and more extensive treatment by Drachmann of the Greek Mechanicians is now in press, 1962.)
5. P. Duhem, *Système du monde,* Vol. I, Paris, 1913.
6. S. Sambursky, *Physics of the Stoics.*
7. F. Solmsen, *Aristotle's System of the Physical World,* Ithaca, 1960.

Chapter 7

See the excellent bibliography for astronomy in Cohen-Drabkin, pp. 563-564. Consult also:

1. O. Neugebauer, *Exact Sciences in Antiquity,* Chap. VI, Appendix I. (A fresh approach to the dependence of Greek astronomy on Babylonian methods.)
2. B. L. van der Waerden, *Die Astronomie der Pythagoreer,* Amsterdam, 1951.
3. A. Diller, "The Ancient Measurements of the Earth," *Isis,* 40 (1949), 6-9.

Chapter 8

There is no separate bibliography for Roman science in Cohen-Drabkin. There are, however, two short accounts with citations of readings:

1. Cyril Bailey, *Legacy of Rome*, Oxford, 1924. (The article on science is by Charles Singer.)
2. P. Brunet and A. Mieli, *Histoire des sciences: Antiquité*, Paris, 1935. (Chaps. 31-33 and 36-38 are on Roman science.) (A new and complete treatment of Roman Science by William Stahl is now in press at the University of Wisconsin, 1962.)

In addition the reader is urged to read the Loeb Library and other modern editions and translations of the principal figures studied in this chapter: Celsus, Lucretius, Vitruvius, Frontinus, Seneca, and Pliny. See also:

3. John Clarke, *Physical Science in the Time of Nero* etc., London, 1910. (Contains a translation of Seneca's *Natural Questions* with notes by the famous geologist Sir Archibald Geike.)
4. Giuseppe Cozzo, *Ingegneria romana*, Rome, 1928.
5. Instituto de Studi Romani, *Le Scienze fisiche e biologiche in Roma e nel Lazio*, Rome, 1933.
6. A. Terquem, *La Science romaine à l'epoque d'Auguste* etc., Paris, 1885. (On Vitruvius.)
7. L. Thorndike, *A History of Magic and Experimental Science*, I, New York, 1923. (Particularly Chaps. 2 and 3, on Pliny and Seneca.)
8. M. Wellmann, *A. Cornelius Celsus*, Berlin, 1913.

PART II

For Part II of our volume, where Cohen-Drabkin is no longer helpful for bibliographies, we give pertinent chapters in the three standard accounts of this period, namely, Duhem's *Système du monde*, cited above in the bibliography of Chapter 6; and the Brunet-Mieli and Thorndike volumes cited in the bibliography for Chapter 8. All these contain bibliographical material to which we shall add a few titles. But these accounts are quite limited in scope and the reader is urged to read the standard editions and translations (when they are available) of the Greek and Latin authors mentioned. For the most part exact citations to the works are given in the numerous quotations throughout Chapters 8 to 13.

Chapters 9 and 10

For the mathematical parts of Chapter 9 use the various works noted in the readings for Chapter 5. For the alchemical materials, see the bibliography in Cohen-Drabkin, pp. 565-566. Consult the following standard treatments:

1. Brunet-Mieli, *Histoire,* Chaps. 44-46 and 48-49.
2. Duhem, *Système,* Vol. 1, Part I, Chap. 5; Vol. 2, Part II, Chap. 1; Vol. 3, Part II, Chap. 3 (Sect. 2).
3. Thorndike, *Magic,* I, Chaps. 4 and 9-23.

Also examine the following more specialized works:

4. C. M. Cochrane, *Christianity and Classical Culture,* Oxford, 1940.
5. F. Cumont, *Astrology and Religion among the Greeks and Romans,* New York, 1912.
6. A. M. Festugière, translator, and A. D. Nock, editor, *Corpus hermeticum,* 2 vols. in 1, Paris, 1945.
7. A. M. Festugière, *La Révélation d'Hermés Trimégiste,* 2 vols., Paris, 1950.
8. H. A. Wolfson, *Philo* etc., 2 vols., Cambridge, Mass., 1947.

For the Christian literature in translation, see:

9. J. B. Lightfoot, editor, *The Apostolic Fathers,* 5 vols., London, 1885-1890.
10. A. Roberts and J. Donaldson, editors, *The Ante-Nicene Fathers,* 10 vols., New York, 1896-1899.
11. P. Schaff, editor, *The Nicene and Post-Nicene Fathers,* 28 vols., New York, 1889-1890.

Chapters 11 and 12

1. Brunet-Mieli, *Histoire,* Chaps. 54-55.
2. Duhem, *Système,* Vol. 3, Part II, Chap. 1, Sects. 1 and 2; Chap. 3, Sects. 1-7.
3. Thorndike, *Magic,* I, Chaps. 27 and 29.

Other specialized works are:

4. H. M. Barrett, *Boethius,* Cambridge, Eng., 1940.
5. E. Brehaut, *An Encyclopedist of the Dark Ages, Isidore of Seville,* New York, 1912.
6. E. S. Duckett, *Gateway to the Middle Ages,* New York, 1938.
7. E. S. Duckett, *Latin Writers of the Fifth Century.* New York, 1930.
8. Erika von Erhardt-Siebold and Rudolf von Erhardt, *The Astronomy of Erigena,* Baltimore, 1940.
9. Erika von Erhardt-Siebold and Rudolph von Erhardt, *The Cosmology in Annotationes in Marcianum: More Light on Erigena's Astronomy,* Baltimore, 1940.
10. C. W. Jones, *Bedae Opera de temporibus,* Cambridge, Mass., 1941. (Introduction has good discussion of *computus* literature.)

11. L. W. Jones, editor, *Cassiodorus' "Introduction to Divine and Human Readings."* New York, 1946. (Has a good introduction.)
12. M. L. W. Laistner, *Thought and Letters in Western Europe* 500-900, London, 1931.
13. H. P. Lattin, "The Eleventh Century Manuscript Munich 14436," *Isis,* Vol. 38 (1948), pp. 205-225. (Contains a wealth of material on early medieval science.)
14. C. E. Lutz, *Johannis Scotti Annotationes in Marcianum,* Cambridge, Mass., 1939.
15. H. R. Patch, *The Tradition of Boethius,* Oxford, 1935.
16. E. K. Rand, *Founders of the Middle Ages,* 2nd ed., Cambridge, Mass., 1929.
17. C. Singer, *From Magic to Science; Essays on the Scientific Twilight,* London, 1928.
18. W. H. Stahl, *Macrobius, Commentary on the Dream of Scipio,* New York, 1952.
19. Thomas Wright, *Popular Treatises on Science Written during the Middle Ages in Anglo-Saxon* etc., London, 1841.

Chapter 13

Some general accounts are:

1. Brunet-Mieli, *Histoire,* Chaps. 47 and 53.
2. Duhem, *Système,* Vol. 1, Chap. 6; Vol. 2, Chap. 9, Sect. 6. (This is the best single account of mechanics in the sixth century.)
3. Thorndike, *Magic,* I.

For Syriac translations, see:

4. A. Baumstark, *Syrische Literatur,* Bonn, 1922.
5. R. Duval, *Littérature syriac,* 3rd ed., Paris, 1907.
6. M. Meyerhof, "New Light on Hunain ibn Ishāq," *Isis,* Vol. 8 (1926), pp. 686-724.
7. W. Wright, *A Short History of Syriac Literature,* London, 1894.

On mechanics, see:

8. A. Haas, "Über die Originalität der physikalischen Lehren des Johannes Philoponus," *Bibliotheca Mathematica,* Dritte Folge, Vol. 6 (1905), pp. 337-342.
9. E. Wohlwill, "Ein Vorgänger Galileis in 6 Jahrhundert," *Physikalische Zeitschrift,* Vol. 7 (1906), pp. 23-32.

On the tradition of classical mathematics in the sixth century, see particularly the introductions to J. L. Heiberg's latest editions of the works of Apollonius and Archimedes.

Index

Note: This index is selective; it does not include references to the further readings assembled on pages 245–51 although it does include the names of authors, editors, and translators mentioned in the body of the volume. It does not include incidental person or place references.

Abelard, 167
Abu 'l-Barakat, 215
Academy of Plato, 46–47, 63, 207
acceleration: Philoponus on, 211–13; Simplicius on, 217; Strato on, 90–92
Aetius of Amide, 219
Agrippa, 129
Albertus Magnus, 171
alchemy, 153–57, 223
Alcmaeon of Croton, 54
Alcuin of York, 203
Alexander, 63
Alexander of Aphrodisias, 92
Alexander of Tralles, 192, 219
Alexandria, Museum at, 47–48, 57
Alfred the Great, 189
algebra: Babylonian, 30–32; Egyptian, 26; Greek, 75–76, 145
Almagest, 119, 143, 222 *and see* Ptolemy, on astronomy
Alpetragius, 215
Ambrose, St., 169, 178
analyses and syntheses, 71–73
Anaxagoras, 52
Anaximander, 49, 53
Anaximenes, 49, 51
Andronicus of Cyrrha, 135
Anthemius of Tralles, 206
Antikythera machine, 124–25
antiperistasis, 213
Apollonius of Perga: *Conics of*, 82–83, 206, 233–35; and epicycles, 118, 119; mentioned, 34, 77, 181, 224
application of areas, 75–76, 234
Apuleius, 191
Aquinas, Thomas, 165
Aratus, 190
Archimedes: on Aristarchus, 115–16; Boethius and, 187, 191; hydrostatics, 97–98; life and works, 78–80; mechanical method, 81–82, 229–31; method of exhaustion, 76, 80, 227–28, 238; reduction to absurdity used by, 72, 227–28; statics, 94, 96–97; works re-edited, 206–7; mentioned, 34, 38, 77, 134, 144, 181, 224
Archytas of Tarentam, 93, 95
Aristaeus, 232
Aristarchus: astronomy of, 114–15; Strato and, 93; trigonometry of, 116
Aristotle: among Latin authors, 138, 192, 194; astronomy of, 110, 113; Chalcidius and, 181; evolution and, 53; general views of nature, 50, 130; laws of motion of, 208; life of, 63–64; logic and logical works of, 38, 187, 191, 220–21; Lyceum of, 47; *Meteorology* of, 151, 154; physical ideas and *Physics* of, 84–89, 172–74, 192, 208–9; scientific method of, 38–41; zoology of, 63–69, 178, 190; mentioned, 37, 42, 71, 137, 195, 224 *et passim.*
Aristotle (pseudo): *Audibles* of, 89; *De mundo*, 138, 191, 221; *De plantis*, 70; *Mechanics* of, 89, 92–95
arithmetic: Babylonian, 28–31; Egyptian, 25–26; Greek, 63, 74–76; *and see* Nicomachus
Assurbanipal, 46
astrology: Greek, 150–52; Plotinus on, 158; Ptolemy and, 144, 152; Tertullian on, 171; mentioned, 22

astronomy: Babylonian, 21–24; Egyptian, 20–22, 24; Greek, Chap. 7 (*passim*); history of early, 64; mentioned, 31, 32
atomic theory: Democritus', 50, 90; Lucretius', 129–33
Augustine, St.: astrology and, 150, 171; Greek philosophy and, 163–64; on matter, 172, 175; on miracles, 170; on scientific accuracy, 167; on time, 172–73
Augustus, 129
Avicenna, 215

Baker, R. H., 107, 108, 122
ballistics, 98
Basil, St.: on animals, 178–79; on astrology, 171; on elements, 173–74; the Hexameron genre and, 169; mentioned, 139, 163
Bede, 169, 197–202
Benedetti, 225
Berthelot, M., 155, 157
blood, system of: Egyptian, 19; Erasistratus', 56, 60–61; Galen's, 60–62; Harvey's, 59; Herophilus', 56; Hippocratic, 55
Boethius: life, works, and translation of, 185–89, 191–92, 221; on sound, 95; mentioned, 144, 181, 183, 193, 203
Bostock, J., 141
botany, 68–70
Boyle, R., 13
Brahe, Tycho, 204
brain: described, 19; dissected, 54
Breasted, J. H., 18–19
Brock, A. J., 59
Bronze Age, 16, 33
Burnet, J., 50

Caesar, J., 128, 129
calendar: Babylonian, 21–24, 25; the *computus* and the, 197–202; Egyptian, 20–21; Julian and Gregorian, 128–29
Callippus, 113
Capella, *see* Martianus
Carmody, F. J., 177–78
Cassiodorus, 181, 187, 189–90, 193, 194
Cassius Felix, 191
Celsus: life and works, 136–37; on medical sects, 57; mentioned, 128, 190
Censorinus, 122
Chalcidius, 181–82, 190
Charlemagne, 202
Chaucer, 189
Childe, V. G., 16
Christianity, 148–49 *and* Chap. 10
Cicero: on astrology, 150, 152; Macrobius and, 183; on Strato, 89; the *Timaeus* and, 181, 190; on the Topics, 187
Clarke, J., 139
Cleanthes, 116
Clement of Alexandria, 165–66
Cleomedes, 117
clocks, 24
Cohen, M. and Drabkin, I. E., 36, 45, 91, 96, 105, 210
computus, 197–202, 205
Contenau, G., 18
Copernicus and Copernican astronomy, 107, 110, 114–15
Cos, 46, 54
Ctesibius, 98, 134, 135, 190

Darwin, 132
decimal systems, 25, 28
Democritus: as an alchemist, 154; his atomism, 50, 90; mentioned, 46
Descartes, 103
Diodorus Siculus, 53
Dionysidorus, 141
Dionysius of Alexandria, 191
Diophantus, 31, 83, 144, 145, 146, 181
diorismos, 74
Dioscorides, 70, 191
dissection, 54, 57–59, 65
Dogmatists, 57, 137
Drabkin, I. E., *see* Cohen, M.
Duhem, P., 93, 169, 174

Ebers, Papyrus, 17
Eleatic school, 37–38, 50
Empedocles, 43, 52
Empiricists, 57, 137
Epicurus, 90, 130
Erasistratus, 56, 57, 60
Eratosthenes, 117–18, 141, 150, 183, 204

Eriugena, John the Scot, 185, 203–4
Ethicus Cosmographicus, 196
Euclid: on the balance, 93, 96; *Elements* of, 74, 77–78; Latin knowledge of, 128, 184, 186, 191–92; life of, 77; *Optics* of, 102; mentioned, 34, 144, 224, 231–32 *et passim*
Eudemus, 64
Eudoxus: astronomy of, 110–13, 216; mathematics of, 76; mentioned, 42, 190, 231
Eutocius, 206–7
evolution, 53
experience and experiment, 16–17, 20, 31, 36–37, 41–45

Festugière, A. M., 150
force: Aristotle on, 87, 208; origin of concept of, 52; Philoponus on, 211–13, 215–16
Freeman, K., 38, 54
Frontinus, 127, 137

Galen: on the blood, 59–62; experimental medicine and, 58–59; translations of, 191, 222; works noted, 144; mentioned, 68, 218, 224
Galileo, 13, 24, 50, 97, 182, 225
Gaudentius, 192
Geminus, 109
geometry: Babylonian, 28, 30; early medieval, 186, 203; Egyptian, 26–27; Greek, Chap. 5, 64; late antique, 145, 146, 160, 206–7; Pythagorean, 36, Chap. 5.
Gerbert, 203
Glover, T. R., 147
Gnosticism, 148
Granger, F., 106
Gregory of Nyssa, 169, 174
Gregory of Tours, 195

Hadrian, 197
Harvey, W., 59
Heath, T. L., 112, 115, 135, 146, 182
Heraclides of Pontus: astronomy of, 114, 135, 182, 184, 200, 204; mentioned, 94
Heraclitus, 49, 51
Hermes and Hermetic literature, 149–50
Hero of Alexandria: on air and vacuum, 43–45; his mathematics, 145; *Mechanics*, 98–101, 145; on optics, 102; Pneumatics of, 101, 136; mentioned, 83, 136, 181, 224
Herodotus, 26
Herophilus, 56, 57
Hicks, R. D., 68
Hindu numerals, 186, 223
Hipparchus: on acceleration, 92, 122–23, 217; astronomy of, 119, 122, 168; on chords, 238
Hippocrates of Chios, 73
Hippocrates of Cos and Hippocratic medicine: concept of "powers" of, 52; corpus, described, 54–56, 62; "healing power of nature" assumed by, 53; *On nature of Man* of, 55; *On Nutriment* of, 55; *Precepts* of, 41; *Sacred Disease* of, 35; mentioned, 43, 58, 191, 218
hippopede, 113
historia, 63, 126
Holmyard, E. J., 153
Homer, 158
Hooke, R., 13
humors, 55
Hunain ibn Ishāq, 222

Iamblichus, 159
inertia, 94, 215
intercalation, 21
Iron Age, 33
irrationality, 75
Isidore of Miletus, 206
Isidore of Seville, St., 141, 184, 193–95
Isis, cult of, 148

Jaeger, W., 47, 63, 64
Jones, C. W., 197
Jones, W. H. S., 35, 41, 55
Jundi-Shapur, 220
Justin, 165
Justinian, 206, 207

Kepler, J., 110
kinematics, 88–89, 91–92
Krusch, Br., 196

Laistner, M. L. W., 203
Lavoisier, A., 51
Leibniz, G. W., 13
Leon of Byzantium, 207
Leucippus, 50
lever, principle or law of: Archimedes on, 96–97, 236–38; Euclid on, 96; Hero on, 100–1; Pseudo-Aristotle on, 93; mentioned, 26
Livingstone, R. W., 60
logic, 37–38; *and see* Aristotle
Lucretius, 129–33
Lyceum: described, 47, 63, 64; physics at, 89–94, Chap. 6; mentioned, 43, 57, 89

Macrobius, 183
magic, 17, 32
Malpighi, M., 59
maps, 29, 129, 144
Marcellus, 79
Martianus Capella: his Nuptials, 183–85, 195; mentioned, 133, 181, 200
mathematics: Babylonian, 28–32, 46, 75; Egyptian, 25–27, 36, 75; Greek, Chap. 6, *et passim*
mechanics, *see* kinematics, motion, Aristotle, Archimedes, *and* Hero
medicine: Egyptian, 17–20; Greek, 54–63, 219
Menaechmus, 231
Menelaus, 145
Menon, 64
method: Aristotelean, 38–41; exhaustion, 76, 80, 227–29; geometric, 71–75; inductive and deductive, 38; Platonic, 50, 84; *and see* Archimedes *and* Aristotle
Metonic cycle, 22
Miletus, 34, 49
motion: Aristotle on, 85–87; laws of, 86, 208–9; Philoponus on, 207–16; Simplicius on, 216–17
Mucianus, 192
mythology, 17, 32, 34

Neo-Platonists and Neo-Pythagoreans, 154–60 (*passim*), 181, 183, 203, 207
Nestorians, 220–21
Neugebauer, O., 9, 23, 30, 115, 122
Newton, I., 13, 52, 77, 83
Nicomachus: *Arithmetic,* 144, 185, 191–92, 194; *Harmonics,* 191, 192; Pythagoreans and, 158–59; mentioned, 83, 184
Nile floods, 64
Nisibis, 220
Nix, L., 100

Ockham, W., 167
Ogle, W., 67
optics, 102–5
Oribasius, 191, 219
Origen, 166, 168
Ostanes, 155

Pappus of Alexandria, 72–73, 74, 83, 146
Parmenides, 37
Paul of Aegina, 218–19
Perin, B., 79
Peutinger Table, 196
Philo of Byzantium, 98, 136
Philo Judaeus, 161–70 (*passim*), 172
Philoponus, John: on *Genesis,* 169, 175; on the *Physics,* 207–17; mentioned, 159, 179
Physiologus, 176–78
pi, 26, 82
Plato: on astronomy, 109, 118, 123; education and, 46; formal view of nature, 50; Neo-Platonists and, 158–60; the *Timaeus* of, 172, 173, 181, 190; mentioned, 38, 39, 63, 68 *et passim*
Pliny: *Natural History* of, 140–41; use of by later authors, 194, 195, 201; mentioned, 128, 138, 181, 191
Plotinus, 158–59
Plutarch, 79, 116
pneumatics, 98, 101
Pompeius Trogus, 190
Porphyry, 187, 221
Posidonius, 118, 150
precession of the equinoxes, 122, 168
Pre-Socratics, 35, 37, Chap. 3 (*passim*)

Priscianus Lydus, 204
Price, D. J. de S., 124, 125
Proclus: career of, 159; on Euclid, 77, 160; on origins of mathematics, 35–36; on reduction, 73; mentioned, 83
Ptolemy, the astronomer: astronomy of, 118–22; on chord lengths, Append. V; *Optics* of, 102–5, 143; on size of earth, 118, 194, translations of, 187, 192, 222; works listed, 143–44; mentioned, 168, 224
Ptolemy, the kings, 48, 77
Pythagoras and Pythagoreans: astronomy of, 106, 109–10, 118; experiments of, 43, 94; mathematics of 75–76, 183, 185, 235 *et passim*; view of nature of, 50; mentioned, 36, 144

Refraction, 103–5
Renatus, 138
Rey, J., 51
Rhind papyrus, 26
Riley, H. T., 141
Rouse, W. H. D., 133
Rufus of Ephesus, 191

Sarton, G., 14
Schiaparelli, G., 110
Sebokt, Severus, 222
Seneca, 138–40, 176, 181
Sergius of Reshaina, 221–22
sexagesimal system, 28–30
Shaw, J. F., 164
Simplicius: on acceleration, 90, 217–18; on Eudoxus' astronomy, 110–12, 216; on weight of air, 216; mentioned, 159
Singer, C., 60
Smith, Edwin, papyrus, 18–20, 31
Snell, W., 103
Solinus, 142, 194
Soranus, 191
Sosigenes, 42
Sothis (Sirius), 21
sound, 43, 94–96
Stoics, 34, 138, 151, 172, 182
Stone Age, 15–16
Strabo, 194
Strato, 43–45, 89–93
surgery, 18–20
Synesius, 155
Syriac translations, 219–23

Temkin, O., 62
Tertullian, 162, 165, 171
Thales, 34, 36, 49, 75
Theodore, 197
Theodosius, 145
Theon of Smyrna, 95, 182
Theophrastus: as associate of Aristotle, 64; his botanical work, 69–70; his will, 47; mentioned, 42, 137, 190, 195, 216
Thomson, D. W., 67
Thorndike, L., 140, 152, 155–57, 171, 194, 219
trigonometry: Aristarchus and, 116; Hipparchus and Ptolemy and, 123–24; Append. V

vacuum and void: Aristotle on, 88, 90; atomic view of, 50, 52, 90, 130; motion in, 130–31, 208–12; Strato on, 43–45, 90
Varro, 128, 133, 184
Vesalius, 225
Vitruvius: on Archimedes, 97–98; on *Architecture* of, 133–36; mentioned, 127, 128, 190

wave theory of sound, 95–96
Webster, E. N., 154
weight, 90, 131 216
Wilson, W., 166
Wolfson, H. A., 161, 170
work, principle of, 98–100